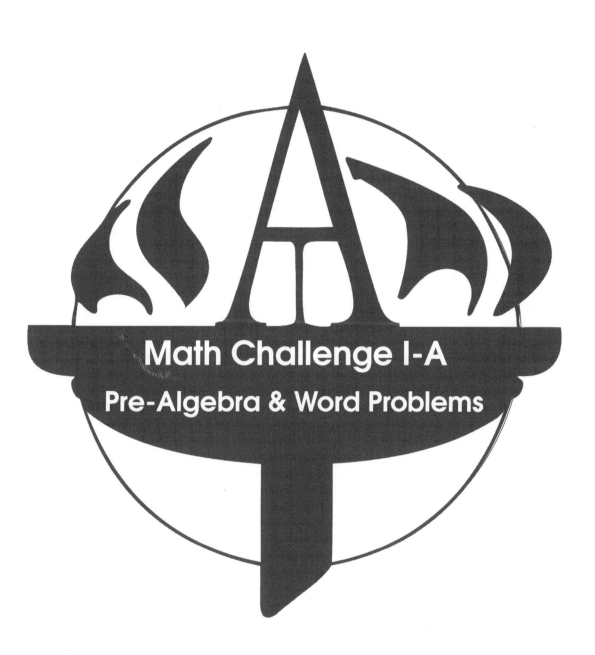

Math Challenge I-A
Pre-Algebra & Word Problems

Areteem Institute

Math Challenge I-A Pre-Algebra & Word Problems

Edited by   David Reynoso
             John Lensmire
             Kevin Wang
             Kelly Ren

ISBN: 1-944863-21-4
ISBN-13: 978-1-944863-21-0

First printing, March 2019.

Math Challenge II-A Combinatorics
Math Challenge II-B Combinatorics
Math Challenge III Combinatorics
Math Challenge I-A Number Theory
Math Challenge I-B Number Theory
Math Challenge I-C Finite Math
Math Challenge II-A Number Theory
Math Challenge II-B Number Theory
Math Challenge III Number Theory

COMING SOON FROM ARETEEM PRESS

Fun Math Problem Solving For Elementary School Vol. 2 (and Solutions Manual)
Counting & Probability for Middle School (and Solutions Manual) - From Common Core to Math Competitions
Number Theory Problem Solving for Middle School (and Solutions Manual) - From Common Core to Math Competitions

The books are available in paperback and eBook formats (including Kindle and other formats).
To order the books, visit https://areteem.org/bookstore.

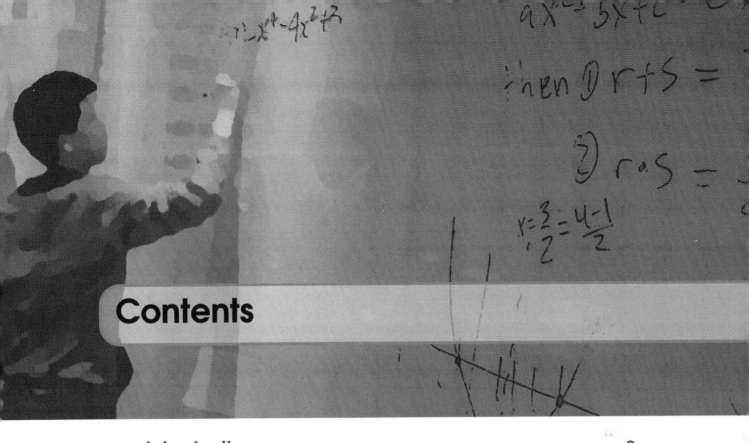

Contents

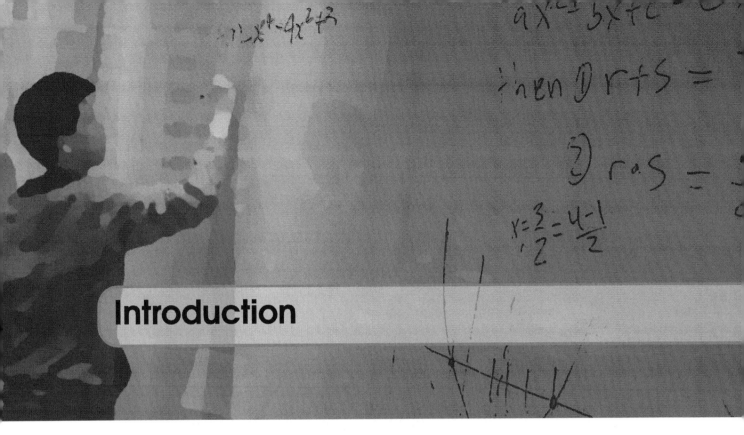

Introduction

Math Challenge I-A is an introductory level course for 6-8 grade students who have no experience in in-depth problem solving nor math competitions. Students learn skills to apply the concepts they learn in school math classes into problem solving. Content includes pre-algebra, fundamental geometry, counting and probability, and basic number theory. Students develop skills in creative thinking, logical reasoning, analytical and problem solving skills. Students are exposed to beginning contests such as AMC 8, MathCounts and Math Olympiads for Elementary and Middle School (MOEMS).

The course is divided into four terms:

- Summer, covering Pre-Algebra and Word Problems
- Fall, covering Geometry
- Winter, covering Counting and Probability
- Spring, covering Number Theory

The book contains course materials for Math Challenge I-A: Pre-Algebra and Word Problems.

We recommend that students take all four terms, but the terms do not build on previous terms, so they do not need to be taken in order and students can take single terms if they want to focus on specific topics.

Students can sign up for the online live or self-paced course at classes.areteem.org.

About Areteem Institute

Areteem Institute is an educational institution that develops and provides in-depth and advanced math and science programs for K-12 (Elementary School, Middle School, and High School) students and teachers. Areteem programs are accredited supplementary programs by the Western Association of Schools and Colleges (WASC). Students may attend the Areteem Institute in one or more of the following options:

- Live and real-time face-to-face online classes with audio, video, interactive online whiteboard, and text chatting capabilities;
- Self-paced classes by watching the recordings of the live classes;
- Short video courses for trending math, science, technology, engineering, English, and social studies topics;
- Summer Intensive Camps held on prestigious university campuses and Winter Boot Camps;
- Practice with selected free daily problems and monthly ZIML competitions at ziml.areteem.org.

Areteem courses are designed and developed by educational experts and industry professionals to bring real world applications into STEM education. The programs are ideal for students who wish to build their mathematical strength in order to excel academically and eventually win in Math Competitions (AMC, AIME, USAMO, IMO, ARML, MathCounts, Math Olympiad, ZIML, and other math leagues and tournaments, etc.), Science Fairs (County Science Fairs, State Science Fairs, national programs like Intel Science and Engineering Fair, etc.) and Science Olympiads, or for students who purely want to enrich their academic lives by taking more challenging courses and developing outstanding analytical, logical, and creative problem solving skills.

Since 2004 Areteem Institute has been teaching with methodology that is highly promoted by the new Common Core State Standards: stressing the conceptual level understanding of the math concepts, problem solving techniques, and solving problems with real world applications. With the guidance from experienced and passionate professors, students are motivated to explore concepts deeper by identifying an interesting problem, researching it, analyzing it, and using a critical thinking approach to come up with multiple solutions.

Thousands of math students who have been trained at Areteem have achieved top honors and earned top awards in major national and international math competitions, including Gold Medalists in the International Math Olympiad (IMO), top winners and qualifiers at the USA Math Olympiad (USAMO/JMO) and AIME, top winners at the

Zoom International Math League (ZIML), and top winners at the MathCounts National Competition. Many Areteem Alumni have graduated from high school and gone on to enter their dream colleges such as MIT, Cal Tech, Harvard, Stanford, Yale, Princeton, U Penn, Harvey Mudd College, UC Berkeley, or UCLA. Those who have graduated from colleges are now playing important roles in their fields of endeavor.

Further information about Areteem Institute, as well as updates and errata of this book, can be found online at http://www.areteem.org.

About Zoom International Math League

The Zoom International Math League (ZIML) has a simple goal: provide a platform for students to build and share their passion for math and other STEM fields with students from around the globe. Started in 2008 as the Southern California Mathematical Olympiad, ZIML has a rich history of past participants who have advanced to top tier colleges and prestigious math competitions, including American Math Competitions, MATHCOUNTS, and the International Math Olympaid.

The ZIML Core Online Programs, most available with a free account at `ziml.areteem.org`, include:

- **Daily Magic Spells:** Provides a problem a day (Monday through Friday) for students to practice, with full solutions available the next day.
- **Weekly Brain Potions:** Provides one problem per week posted in the online discussion forum at `ziml.areteem.org`. Usually the problem does not have a simple answer, and students can join the discussion to share their thoughts regarding the scenarios described in the problem, explore the math concepts behind the problem, give solutions, and also ask further questions.
- **Monthly Contests:** The ZIML Monthly Contests are held the first weekend of each month during the school year (October through June). Students can compete in one of 5 divisions to test their knowledge and determine their strengths and weaknesses, with winners announced after the competition.
- **Math Competition Practice:** The Practice page contains sample ZIML contests and an archive of AMC-series tests for online practice. The practices simulate the real contest environment with time-limits of the contests automatically controlled by the server.
- **Online Discussion Forum:** The Online Discussion Forum is open for any comments and questions. Other discussions, such as hard Daily Magic Spells or the Weekly Brain Potions are also posted here.

These programs encourage students to participate consistently, so they can track their progress and improvement each year.

In addition to the online programs, ZIML also hosts onsite Local Tournaments and Workshops in various locations in the United States. Each summer, there are onsite ZIML Competitions at held at Areteem Summer Programs, including the National ZIML Convention, which is a two day convention with one day of workshops and one day of competition.

ZIML Monthly Contests are organized into five divisions ranging from upper elementary school to advanced material based on high school math.

- **Varsity:** This is the top division. It covers material on the level of the last 10 questions on the AMC 12 and AIME level. This division is open to all age levels.
- **Junior Varsity:** This is the second highest competition division. It covers material at the AMC 10/12 level and State and National MathCounts level. This division is open to all age levels.
- **Division H:** This division focuses on material from a standard high school curriculum. It covers topics up to and including pre-calculus. This division will serve as excellent practice for students preparing for the math portions of the SAT or ACT. This division is open to all age levels.
- **Division M:** This division focuses on problem solving using math concepts from a standard middle school math curriculum. It covers material at the level of AMC 8 and School or Chapter MathCounts. This division is open to all students who have not started grade 9.
- **Division E:** This division focuses on advanced problem solving with mathematical concepts from upper elementary school. It covers material at a level comparable to MOEMS Division E. This division is open to all students who have not started grade 6.

To participate in the ZIML Online Programs, create a free account at `ziml.areteem.org`. The ZIML site features are also provided on the ZIML Mobile App, which is available for download from Apple's App Store and Google Play Store.

Acknowledgments

This book contains many years of collaborative work by the staff of Areteem Institute. This book could not have existed without their efforts. Huge thanks go to the Areteem staff for their contributions!

The examples and problems in this book were either created by the Areteem staff or adapted from various sources, including other books and online resources. Especially, some good problems from previous math competitions and contests such as AMC, AIME, ARML, MATHCOUNTS, and ZIML are chosen as examples to illustrate concepts or problem-solving techniques. The original resources are credited whenever possible. However, it is not practical to list all such resources. We extend our gratitude to the original authors of all these resources.

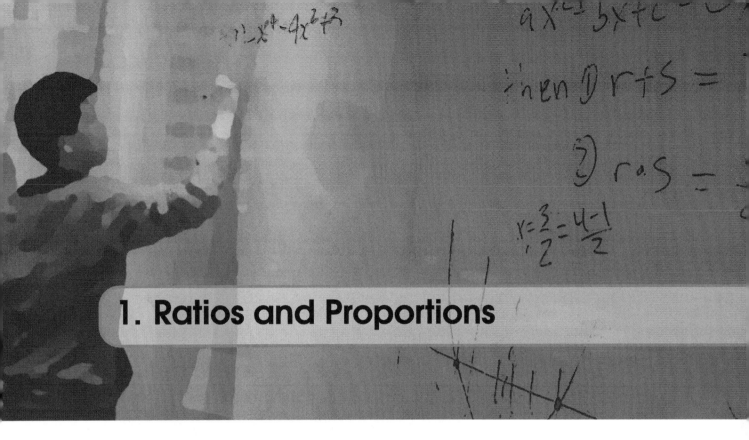

1. Ratios and Proportions

1.1 Example Questions

Problem 1.1 Find x in the following proportions.

(a) $12 : 17 = x : 68$.

(b) $1.68 : 12 = 7 : x$.

(c) $95 : 6.5 = x : 5.2$

Problem 1.2 Solve the following involving ratios.

(a) Thomas mixed 3 pints of red paint with 4 pints of blue paint to make a new color. He now wants to use 27 pints of red paint and some blue paint to make the same color. How many pints of blue paint will he need?

(b) The ratio of boys to girls in Math Zoom Academy is 5 : 3. If there are 24 more boys than girls in Math Zoom Academy, then how many girls are in Math Zoom Academy?

Problem 1.3 Solve the following questions involving percentages.

(a) The table shows some of the results of a survey. What percentage of the males surveyed read the newspaper?

	Read	Don't Read	Total
Male	?	26	?
Female	58	?	96
Total	136	64	200

(b) The table below gives the percent of students in each grade at school A and school B.

	K	1	2	3	4	5	6
A	21%	12%	11%	15%	13%	17%	11%
B	18%	11%	16%	11%	13%	14%	17%

School A has 100 students, and school B has 200 students. If the two schools combined, what percent of the students are in grade 6?

Problem 1.4 Percent Increase and Decrease

(a) A merchant offers a large group of items at 30% off. Later, the merchant takes 20% off these sale prices and claims that the final price of these items is 50% off the original price. What is the true total discount?

(b) The number of students going on a field trip changed from 24 to 36. What was the percentage increase in the number of students going on a field trip? Later, the number of students going changed back from 36 to 24. What was the percentage decrease in the number? Give both answers as a percentage rounded to the nearest whole number.

Problem 1.5 Discounts, Coupons, and Sales Tax

(a) Samantha wanted to buy a new iPhone case that was originally priced $28, and she had a coupon for 15% off. What would be the discounted price of the iPhone case, after the coupon was applied?

(b) In the same situation as part (a), if the state tax rate was 8%, what was the final amount Samantha had to pay?

(c) The cashier, Roger, applied the tax first and got the amount Samantha needed to pay before Samantha showed him the coupon. Roger took the coupon and applied the 15% discount and reached a final amount for Samantha to pay. Samantha made the payment and received her new lemon-patterned iPhone case. She was very happy and put it on the phone immediately. However, on her way home, she kept wondering if she might have ended up paying more tax because the cashier applied the tax before applying the coupon. Can you help her figure it out?

Problem 1.6 When gold sold for $16 an ounce, Johnny found $6 worth of gold in his claim. Gold presently sells for $328 an ounce. How many dollars is Johnny's amount of gold worth today?

Problem 1.7 John is on a business trip with his friends Eric, Nick and Oscar. Oscar lost all his money, so the friends wanted to help him. Each gave Oscar the same amount of money. However, Eric gave Oscar 20% of his own money, John gave Oscar 25% of his money, and Nick gave Oscar $\frac{1}{3}$ of his money. What percent of the group's money does Oscar now have?

Problem 1.8 Henry reads 160 pages of a book per day. After 5 days, Henry has $\frac{3}{5}$ of the book remaining. How many pages does the book have?

Problem 1.9 Sixty percent of the people on a subway train are seated. As some people prefer standing, only 75% of the seats on the subway are filled. If there are 12 empty seats, how many people are on the train?

Problem 1.10 A fruit salad consists of crabapples, cranberries, black berries, and black cherries. If there are twice as many cranberries as crabapples and three times as many black berries as black cherries and four times as many black cherries as cranberries and the fruit salad has 280 total fruits, then how many black cherries does it have?

1.2 Quick Response Questions

Problem 1.11 Last month the price of gas was $1.10 per gallon. This month gas is selling for $1.32 per gallon. Find the percentage increase in the price of gas per gallon.

Problem 1.12 There are 180 days in a school year. John was present 85% of the total days. How many days was he present?

Problem 1.13 Jim receives a weekly salary of $200. He spends $60 per week on gas. What percent of his weekly salary does he use for gas?

Problem 1.14 A coffee mug marked $15.00 was on sale for $12.00. What is the rate of discount as a percent?

Problem 1.15 The ratio of girls to boys participating in intramural volleyball at Ashland Middle School is 7 to 4. There are 42 girls in the program. What is the total number of participants?

Problem 1.16 Cathy and Jimmy looked for seashells at Newport Beach. For every 9 seashells Cathy found, Jimmy found 7. Cathy found 54 seashells. How many fewer seashells did Jimmy find than Cathy?

Problem 1.17 Carson bought five notebooks from the College Bookstore at a cost of $2.50 each. His brother Derick liked the notebooks and went to the bookstore the following day and also bought 5 notebooks. The bookstore had a 20%-off sale that day. How much did Derick save (in dollars) compared to Carson on the purchase of the five notebooks?

Problem 1.18 Jess paid $6.31 for a shirt marked 25% off the regular price. What was the regular price of the shirt in dollars?

Problem 1.19 Samantha has exactly 2 pennies, 2 nickels, one dime and one quarter in her pocket. What percent of a dollar is in her pocket?

Problem 1.20 A $480 TV was put on sale for 30% off. It wasn't sold so the price was lowered an additional percent off the sale price making the new price $285.60. What percentage was the second discount?

1.3 Practice Questions

Problem 1.21 Are the following ratios proportional?
(a) 3 : 4 and 24 : 32

(b) 2 : 3 and 4 : 9

(c) 7 : 9 and 28 : 36

Problem 1.22 The ratio of boys to girls at the baseball game is 5 : 2. There are 28 girls. How many more boys are there than girls?

Problem 1.23 Calvin bought four Avengers movie tickets for his friends at a cost of $12.50 each. His friend Mark wanted to watch the movie with his family as well and went to buy the same amount of tickets the following day. The theater had a 20%-off sale that day. How much did Mark save comparing to Calvin on the purchase of four movie tickets?

Problem 1.24 The sales tax rate in Orange County is 8%. During a sale at an outlet in Orange County, the price of a suit is discounted 40% from its $190.00 price. Two clerks, Jimmy and Tony, calculate the bill independently. Jimmy first adds 8% sales tax to the price, and then subtracts 40% from this total. Tony first subtracts 40% of the price from the original price, and then adds 8% sales tax to the discounted price. What is Jimmy's total minus Tony's total?

Problem 1.25 Conner was taking an investment course at a local community college. In the course he was given a practice account with imaginary currency, and a list of imaginary companies for investment. During the first class period, he invested equal amounts from the account on stocks of three of the imaginary companies AFZ, BLS, and CIW. The AFZ stock never changed value for the entire course. The BLS stock increased by 10% by the end of the second class, and then decreased by 10% of the new value by the end of the third class. The CIW stock decreased by 20% by the end of the second class, and then increased by 20% of the new value by the end of the third class. At the end of the third class, which stock had the highest value, and which had the lowest?

Problem 1.26 A collector offers to buy the 1967 year of the sheep stamp sheet for 2000% of its face value. Bridget has one of the sheets with 12 stamps that had an original face value of 25 cents per stamp. How much would Bridget receive if she sold it to the collector?

Problem 1.27 Andy had no money, so his Granny Smith gave him 36% of her money. Now Granny Smith has $80 left, and Andy has $2 more than Elberta. How many dollars does Elberta have?

Problem 1.28 Jenny starts with a full jar of jellybeans. Each day, she eats 20% of the jellybeans that were originally in the jar. At the end of the second day, 36 jellybeans remain. How many jellybeans were in the jar originally?

Problem 1.29 Two-thirds of the monkeys in a cage are seated in three-fourths of the spots. The rest of the monkeys are standing. If there are 6 empty spots, how many monkeys are in the cage?

Problem 1.30 John, Edward, and Dan did a fundraiser for the math club at school and raised a total of $370. They divided the $370 into three parts such that the second part is $\frac{1}{4}$ of the third part and the ratio between the first and the third part is $3 : 5$. Find the value of each part.

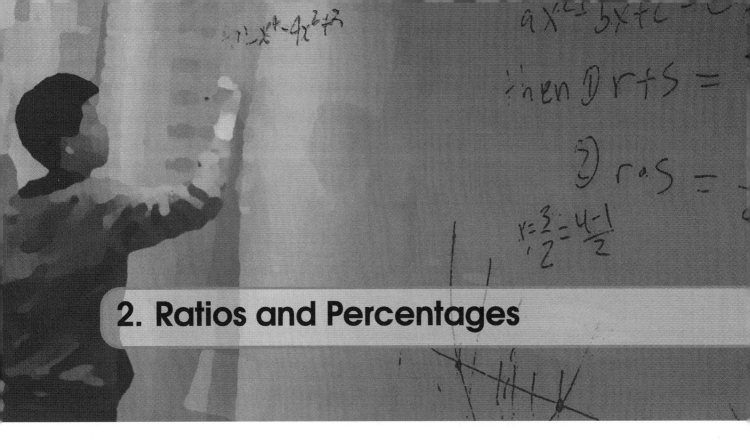

2. Ratios and Percentages

2.1 Example Questions

Problem 2.1 Ratios warmup problems!

(a) In a far-off land three fish can be traded for two loaves of bread and one loaf of bread can be traded for six ears of corn. How many ears of corn are worth the same as one fish?

(b) The ratio of llamas to ostriches in the Math Zoom Academy petting zoo is $4 : 7$. If there are total of 44 llamas and ostriches in the petting zoo, how many of the them are llamas?

Problem 2.2 Use percentages to solve the following problems.

(a) Three bags of jelly beans contain 26, 28, and 30 beans. The bags consist of respectively 50%, 25%, and 20% yellow jelly beans. All three bags of beans are dumped into one bowl. What percent of all beans are yellow jelly beans? Round your answer to the nearest percent.

(b) Jong-Zhi took a math test that had 12 arithmetic questions, 15 algebra questions and 18 geometry questions. She got 75% of the arithmetic questions correct and 60% of the algebra questions correct. How many of the geometry questions must she get correct to get a passing grade of 75%?

Problem 2.3 Business is a little slow at Lou's Fine Shoes, so Lou decides to have a sale. On Friday, Lou increases all of Thursday's prices by 10%. Over the weekend, Lou advertises the sale: "Ten percent off the list price. Sale starts Monday." How much does one pair of shoes cost on Monday that cost $40 on Thursday?

Problem 2.4 In the popular TV show "Who Wants to be a Millionaire", contestants earn certain amount of money based on the number of questions they answer correctly. The dollar values of each question are shown in the following table (where k = 1000).

Question	1	2	3	4	5	6	7	8
Value	100	200	300	500	1k	2k	4k	8k
Question	9	10	11	12	13	14	15	
Value	16K	32K	64K	125K	250K	500K	1000K	

Between which two questions is the percentage increase of the value the smallest?

Problem 2.5 Jim is paid an 8% commission on the first $800 of weekly sales, and a 14% commission on any sales past $800. If Jim's sales were $1300, what was his commission?

Problem 2.6 Vitamin tablets are packed in three different sized bottles: small (S), medium (M) and large (L). The medium size costs 50% more than the small size and contains 20% fewer tablets than the large size. The large size contains twice as many tablets as the small size and costs 30% more than the medium size. Rank the three sizes from best to worst buy.

Problem 2.7 Two 600 ml pitchers contain orange juice. One pitcher is 30% full and the other pitcher is 40% full. Water is added to fill each pitcher completely, then both pitchers are poured into one large container. What percent of the mixture in the large container is orange juice?

Problem 2.8 Phil Lanthropist won a great deal of money in a contest. He gave 20% of his winnings to his parents, gave 25% of the remaining money to his children, and gave the remaining $900,000 to his favorite charity. What was the total number of dollars that Phil won?

Problem 2.9 At a party there are only single women and married men with their wives. 40% of the women are single. What percentage of the people in the room are married men?

Problem 2.10 In the fish tank at Albert's house, $\frac{1}{4}$ of the fish are red and the number of black fish is $\frac{3}{5}$ of the number of red fish. There are 24 additional fish that are all spotted. How many red fish are there?

2.2   Quick Response Questions

Problem 2.11 There are 2 packs of crayons available for every 5 students at Amy's art class. How many students can share 18 packs of crayons?

Problem 2.12 If 5 bananas are worth the same as 3 apples then how many bananas are worth the same as 15 apples?

Problem 2.13 Sam received a grade of 80% on a geometry test. If he solved 24 problems correctly, how many problems were on the test?

Problem 2.14 A salesman who works on a commission basis earns 18% of his sales. How many dollars was his commission on a $480 sale?

Problem 2.15 A clerk at the stationary store receives $12\frac{1}{2}$% commission on all merchandise sold. If she received $52 in commission last week, what were her sales for the week in dollars?

Problem 2.16 Last year Areteem Institute had an enrollment of 500 students. This year the enrollment is 800 students. What is the percent of increase in student enrollment?

Problem 2.17 The ratio of the length of Mary's cat to the length of Amy's cat is 5 : 7. Mary's cat is 100 cm long. How much longer is Amy's cat than Mary's cat in cm?

Problem 2.18 John bought a coat which usually sells for $98.00 at 25% off. How many dollars he pay for the coat?

Problem 2.19 An agent receives a commission of 6% of the selling price of a house. The rest of the proceeds go to the owner of the house. If the agent sells a house for $135,000, how much does the house owner receive?

Problem 2.20 Sally is playing basketball. After Sally takes 20 shots, she has made 55% of her shots. She takes 5 more shots and she raises her percentage to 56%. How many of the last 5 shots did she make?

2.3 Practice Questions

Problem 2.21 Use ratios to solve the following problems!

(a) A middle school has 780 students, some of which go to Math Olympiad classes. Among those who attend Math Olympiad classes, $\frac{8}{17}$ are in 6th grade, and $\frac{9}{23}$ are in 7th grade. How many students do NOT attend Math Olympiad classes?

(b) Rita has 36 marbles, 20 of which are red and 16 of which are white. Rose has 27 marbles, all of them either red or white. Suppose Rita and Rose have the same ratio of red to white marbles. How many more white marbles does Rita have than Rose?

Problem 2.22 In her history class, Marie averaged 90% correct on five 10 question quizzes, got 96% correct on a 50 question midterm exam, and 75% correct on an 80 question final exam. What is the percentage of correct answers she provided if the total points for correct answers on all quizzes and exams are combined?

Problem 2.23 A shopper buys a $100 coat on sale for 20% off. An additional $5 is taken off the sale price by using a discount coupon. A sales tax of 8% is paid on the final selling price. What is the total amount the shopper pays for the coat?

Problem 2.24 Chris' pay went from $20 per hr to $25 per hr after his first evaluation. After his second evaluation his pay was raised to $33 per hr. What is the difference between the second raise as a percent and the first raise as as percent?

Problem 2.25 Linda receives a weekly salary of $120 plus a commission of 5% on all sales above $500 per week. During three weeks Linda's total sales were $1540, $1235, and $1040. What was her total paycheck for the three weeks?

Problem 2.26 A grocery store sells eggs in three sizes: small (S), medium (M) and large (L). The medium size costs 50% more than the small size and contains 20% fewer eggs than the large size. The large size contains twice as many eggs as the small size and costs 30% more than the medium size. Rank the three sizes from best to worst buy.

Problem 2.27 A and B are two identical cups. A is full with salt water containing 2% salt and B is half full with salt water containing 0.8% salt. Suppose we pour half of the salt water from cup A to cup B so cup B is now full of salt water. What percentage of salt is the salt water in cup B?

Problem 2.28 Sally baked 5 dozen cookies on Saturday afternoon. She gave 60% of the cookies to her neighbors at the neighborhood barbecue. On Sunday, she took 75% of the remaining cookies to the church social. On Monday night, she and her family ate 50% of the remaining cookies while watching football. What percent of the 5 dozen cookies remain?

Problem 2.29 The students in Miss Einstein's class took a math test. Two-thirds of the class passed and the other third failed. The ratio of boys to girls in the class is 2 to 1. All of the girls passed the exam. What percentage of boys failed the exam?

Problem 2.30 A bag contains 3 blue, 4 red and 3 yellow marbles. How many blue marbles must be added to the bag for it to contain 75% blue marbles?

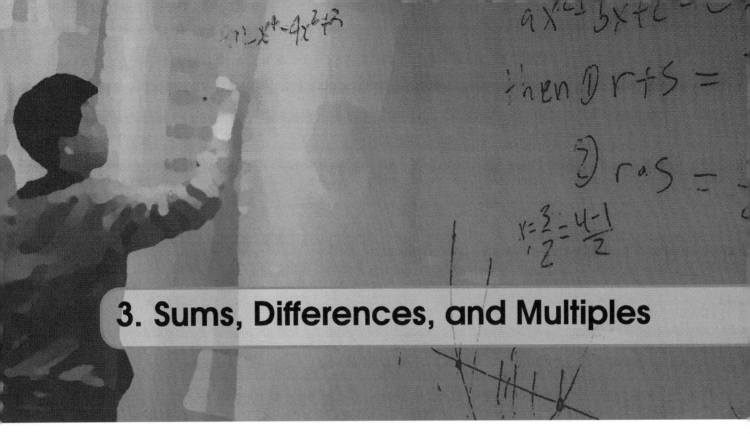

3. Sums, Differences, and Multiples

3.1 Example Questions

Problem 3.1 Solve the following.
(a) A school bought some basketballs and volleyballs, 50 balls in total. There are 10 more basketballs than volleyballs. How many balls of each kind are there?

(b) When Matt was 17 years old, Nathan was 23. This year the sum of their ages is 50. What are their ages this year?

(c) David and Ed went to pick cherries. Together they picked 82 cherries. If David gave 4 cherries to Ed, they would have the same number of cherries. How many cherries did each of them pick?

Problem 3.2 Solve the following problems.

(a) Alice, Beth, and Cynthia have $50 in total. Alice has twice as much money as Beth, and Beth has 3 times as much as Cynthia. How much money does each have?

(b) Allison is 14 years old, and her dad is 50 years old. How many years ago was Allison's dad's age 5 times Allison's age?

Problem 3.3 Answer the following.

(a) In the orchard there are 144 trees, which are either apple trees or peach trees. If there were 12 fewer apple trees and 20 more peach trees, then the numbers of the two kinds of trees would have been the same. How many trees of each kind are there?

(b) Mary, Rose and Catherine are working on a school project and have some big pieces of ribbon of different sizes. If they glue together their pieces of ribbon they get a ribbon that is 840 centimeters long. Their teacher measured each of their ribbons and told them that Rose's ribbon is 130 centimeters longer than Mary's, and Catherine's is 220 centimeters longer than Rose's. Can you figure out how long is each ribbon?

Problem 3.4 In a school, there are 165 students in the third and fourth grades altogether. If there were 6 more third graders, the third grade would have twice as many students as the fourth grade. How many students are there in each of the two grades?

Problem 3.5 Coach just brought a *huge* bag with 75 balls for gym class. He told us that in the bag there are twice as many basketballs as soccer balls, and that there are 3 more volleyballs than soccer balls. He won't let us play with them until we figure out exactly how many of each he brought. Can you help us out?

Problem 3.6 I just found a box with Christmas ornaments in the attic. It has a huge label on the front that says it contains 200 Christmas ornaments of three different colors (red, blue, and white). I'm so clumsy that I broke some of them by accident. It seems I broke 14 white ornaments and now I have the same number of white ornaments and blue ornaments. Mom says that before I broke them we had 4 more red ornaments than 3 times the number of white ornaments. How many red ornaments were there in the box before I found it?

Problem 3.7 David has 3 times as much money as Chris. If David spends $240, and Chris spends $40 dollars, they will have the same amount of money. How much money do each of them have originally?

Problem 3.8 The average price of a basketball, a soccer ball, and a volleyball is $36. The basketball is $10 more expensive than the volleyball, the soccer ball is $8 more expensive than the volleyball. How much is the soccer ball?

Problem 3.9 Captain Hook, Popeye, and Sinbad went out to the sea to hunt for treasure. They all found diamonds. The total number of diamonds found by Captain Hook and Popeye is 80. The total number of diamonds found by Popeye and Sinbad is 70. The total number of diamonds found by Captain Hook and Sinbad is 50. How many diamonds did each of them find?

Problem 3.10 In the morning, there were 149 vehicles (cars and trucks) in the parking lot. In the afternoon 8 more cars entered the parking lot, and 5 trucks drove away. It turns out that 3 times the number of cars in the parking lot is the same as 5 times the number of trucks. How many cars and how many trucks were there in the parking lot in the morning?

3.2 Quick Response Questions

Problem 3.11 Aiden and Brian have 12 coins altogether. Aiden has twice as many coins as Brian. How many coins does Brian have?

Problem 3.12 Patrick and Joe ate all the cookies that their mom had just made. If her mom made 22 cookies and Joe ate 10 more cookies than Patrick. How many cookies did Patrick eat?

Problem 3.13 David likes collecting his bus tickets. Last week he took the bus 6 times less than this week and he gathered 34 bus tickets altogether. How many times did he ride the bus this week?

Problem 3.14 Troy bought a juice box for 25 cents and a burger for 5 times as much as the juicebox. How many dollars did Troy spend?

Problem 3.15 Tom and Jerry went fishing. They caught 60 fish altogether. Tom caught 3 times as many fish as Jerry. How many fish did Jerry catch?

Problem 3.16 There are 65 oranges in a grocery store arranged in two piles. If the first pile has 5 more oranges than the second pile, how many oranges does the small pile have?

Problem 3.17 David's dad bought a digital camera from an online store. He paid the price of the camera plus shipping and handling, for a total of $270. The price is 8 times the cost of shipping and handling. What is the price in dollars of the digital camera?

Problem 3.18 A country sent 108 athletes to the Olympic Games, among which the number of male athletes is twice the number of female athletes. How female athletes are there?

Problem 3.19 Aiden and Brandon collected 69 rare coins altogether. Aiden collected 2 times more coins than Brandon. How many coins did Brandon collect?

Problem 3.20 Alek and Abby are collecting toy cars. Together they have 35 toy cars and Abby has 7 less than Alek. How many toy cars does Alek have?

3.3 Practice Questions

Problem 3.21 Two hungry ant-eaters were feeding from a small anthill. If one of the ant-eaters eats 150 more ants than the other and they ate a total of 2300 ants, how many did each of them eat?

Problem 3.22 Your best friend is having a birthday party. Since his twin sisters' birthday is so close, their parents decided to have one party for the three of them at the same time. They ask you for help placing the candles on each of their cakes, but you can't remember how old each of them are becoming this year. They give you a total of 20 candles, and you also know that this year the age of your friend will be 3 times the age his twin sisters. How many candles should you place in each cake?

Problem 3.23 In a school there are totally 108 students in the third, fourth, and fifth grades. The third grade has 11 fewer students than the fourth grade, and the fourth grade has 16 more students than the fifth grade. How many students are there in each grade?

Problem 3.24 Tom and Jerry went fishing together and each caught some fish. If Tom gave 2 fish to Jerry, then they would have the same number of fish. If Jerry gave 2 fish to Tom, then the number of Tom's fish would be 5 times that of Jerry's. How many fish did each of them catch?

Problem 3.25 The school bought some basketballs, volleyballs, and soccer balls. There are 40 more soccer balls than volleyballs, and 8 fewer basketballs than volleyballs. Also, the number of soccer balls is 4 times the number of basketballs. How many balls of each kind are there?

Problem 3.26 The sum of the ages of Lisa and Suzi is 24. Four years ago Lisa's age was three times Suzi's age. What are their current ages?

Problem 3.27 There are 48 students in a class. If 3 more boys joined the class, the number of boys would be twice the number of girls. What is the current number of boys in the class?

Problem 3.28 John, Jack, and Jill have 159 marbles altogether. John has 2 more marbles than Jack, and if Jill gave 5 marbles to Jack, Jack would have the same number of marbles as Jill. How many marbles does each of them have?

Problem 3.29 Old McDonald and Old Wendy went to the market to sell apples. They originally had the same amount of apples. Old McDonald sold 11 pounds of apples, and Old Wendy sold 29 pounds. Now the remaining amount of apples of Old McDonald is 3 times as many as the amount of apples of Old Wendy. How many pounds of apples do each of them have now?

Problem 3.30 A baseball game is being played. Some people are watching at the stadium and some are watching on TV at home. The number of people who watch on TV is 480 more than the number of people in the stadium. If 50 people at the stadium went home and watched the game on TV, the number of people who watch on TV would be 5 times the number of the number of people in the stadium. How many people are watching the game in total?

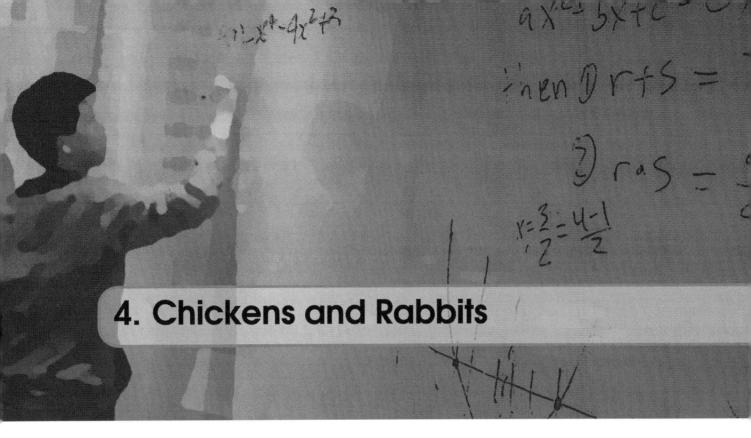

4. Chickens and Rabbits

4.1 Example Questions

Problem 4.1 There are some chickens and some rabbits on a farm. Suppose there are 50 heads and 140 feet in total among the chickens and rabbits, how many chickens are there? How many rabbits?

Problem 4.2 Answer the following using the Chicken and Rabbit Method

(a) Mary is baking a lot of cookies for a bake sale at Samantha's school. To bake the cookies she needs a total of 60 eggs. The eggs come in small cartons containing 6 eggs or large cartons containing 12 eggs. If Mary buys a total of 7 cartons, how many small cartons and how many large cartons does she buy?

(b) Morgan needed 70 sticks for a project at school. Each stick is either 3 inches or 5 inches and the total length of all the sticks combined is 270 inches. How many 3 inch sticks and 5 inch sticks are there?

Problem 4.3 The counselor brought his 51 students to the lake to go rowing, 6 people for each big boat and 4 people for each small boat. They rented 11 boats to fit everyone with no empty seats. How many big and small boats each did they rent?

Problem 4.4 Sasha takes a mathematics competition. There are a total of 20 problems. For each correct answer, competitors receive 5 points. For each wrong answer, they instead get 1 point taken away. Sasha has 64 points in total. How many problems did she answer correctly?

Problem 4.5 Solve the following.

(a) Teachers and students from the Areteem Summer Camp visited the museum. They bought a total of 99 tickets for 218 dollars. If each teacher ticket costs 4 dollars, and each student ticket costs 2 dollars, how many teachers and students were there respectively?

(b) A group of 68 people rent 24 motorcycles of two kinds at a racetrack. The first kind has a capacity of 2 people and costs \$40 per motorcycle. The second has a capacity of 3 people and costs \$30 per motorcycle. The 68 people exactly fill all vehicles. What is the total cost in renting the 24 motorcycles?

Problem 4.6 Jack went hiking. His uphill speed was 3 miles per hour and downhill speed was 5 miles per hour, and he hiked a total of 6 hours including both uphill and downhill, with total distance 23 miles. How many hours did he spent uphill and downhill respectively?

Problem 4.7 Answer the following questions.

(a) There are some chickens and some rabbits on a farm. Suppose there are 50 heads and there are 20 more rabbit feet than chicken feet on the farm. How many chickens and how many rabbits are there?

(b) The Math Club collected donations from 40 people who live in either City A or B. Each person from City A contributed $5, and each person from City B contributed $8. In total, $5 more was collected from City A than from City B. How many people are there in each city?

Problem 4.8 More Practice with Differences.

(a) Four basketballs and five volleyballs cost 185 dollars in total. If a basketball costs 8 dollars more than a volleyball, what is the cost of one basketball?

(b) It requires either 45 small trucks or 36 big trucks to transport a batch of steel blocks. Given that each big truck can load 4 more tons of steel blocks than each small truck. How many tons of steel blocks are in a batch?

Problem 4.9 100 monks eat 100 steamed buns. If each senior monk eats 4 steamed buns, and 4 junior monks eat 1 steamed bun, how many senior monks and junior monks are there?

Problem 4.10 Suppose there are chicken, rabbits, and sheep on a farm. There are 70 heads in total and 220 feet. If there are the same number of rabbits and sheep, how many chickens, rabbits, and sheep are on the farm?

4.2   Quick Response Questions

Problem 4.11 A farmer has 40 chickens and 20 rabbits on a farm. How many legs are on the farm?

Problem 4.12 There are 50 motorcycles and 40 cars parked in a parking lot. Terry counts the number of motorcycle wheels and Jerry counts the number of car wheels. How many more wheels does Jerry count than Terry?

Problem 4.13 In a farm there are goats and ducks. The total number of heads is 100, and the total number of legs is 316. How many goats are there?

Problem 4.14 A pet owner has cats and birds. There are 25 pets in total and all combined the pets have a total of 90 legs. How many cats are there?

Problem 4.15 Sarah counts her chickens and rabbits, and there are 16 heads and 44 feet. How many rabbits are there?

Problem 4.16 The price of a pack of colored pencils is $19 and the price of a pack of regular pencils is $11. The math teacher bought 16 packs of pencils for a total of $280. How many packs of colored pencils did the math teacher buy?

Problem 4.17 Two trucks dump dirt of 400 cubic meters. Truck A carries 7 cubic meters per load. Truck B carries 4 cubic meters per load. The dirt is removed after 70 loads. How many loads are carried by truck A?

Problem 4.18 In a math competition, there are 25 questions. Each correct answer earns 6 points. One point is taken away for each incorrectly answered or unanswered question. Jenny received 101 points. How many questions did she answer correctly?

Problem 4.19 Aria has 16 coins that are nickels and dimes. The total value is \$1.20. How many dimes does Aria have?

Problem 4.20 In a farm the total number of chickens and rabbits is 100. If the number of chicken feet is 80 more than the number of rabbit feet, how many chickens are there?

4.3   Practice Questions

Problem 4.21 There are some chickens and some rabbits on a farm. Suppose there are 45 heads and 128 feet in total among the chickens and rabbits, how many of the animals are chickens? How many are rabbits?

Problem 4.22 Sixty vehicles (cars and motorcycles) are parked in a parking lot. Totally there are 190 wheels. Given that a car has 4 wheels and a motorcycle has 2 wheels, how many cars and motorcycles each are in the parking lot?

Problem 4.23 Use 400 matches to make triangles and pentagons. Totally 88 triangles and pentagons are made with no matches left over. How many of each shape are made?

Problem 4.24 There are 20 questions in a math competition. Five points are given to each correct answer, and -3 points are for each incorrect answer or unanswered question. Jeff received 60 points in the competition. How many questions did he answer correctly?

Problem 4.25 A large family of 20 people goes to a restaurant. They each order either pizza or salad. The pizza costs \$9.00 and salad costs \$3.00. In all the family spends \$138.00. How many pizzas and how many salads did the family order?

Problem 4.26 Ricky is on a long road trip. He wants to drive 390 miles in the next 8 hours. Some of the time he will be able to drive 60 miles per hour, and some of the time he will have to drive slower at 30 miles per hour. If Ricky accomplishes his goal and drives exactly 390 miles in 8 hours, how many hours does he drive at 60 miles per hour?

Problem 4.27 There are many ducks and sheep in a farm. If we count the heads, there are total of 80 heads. If we count the legs, there are 56 more legs from sheep than from ducks. How many ducks and how many sheep are there in the farm.

Problem 4.28 Patrick and George both work gardening over the summer. Patrick has more experience, so he earns $5 more per hour than George. Patrick works 10 hours, and George works 20 hours, and they earn a combined total of $410. How much does Patrick earn per hour?

Problem 4.29 100 mice eat 100 cakes. If each big mouse eats 3 cakes, and 3 baby mice eat 1 cake, how many big mice and baby mice are there?

Problem 4.30 Tony's mom took out $380 from the bank. There are $2, $5, and $10 bills, a total of 80 bills. The number of $5 bills and $10 bills are the same. How many bills of each type are there?

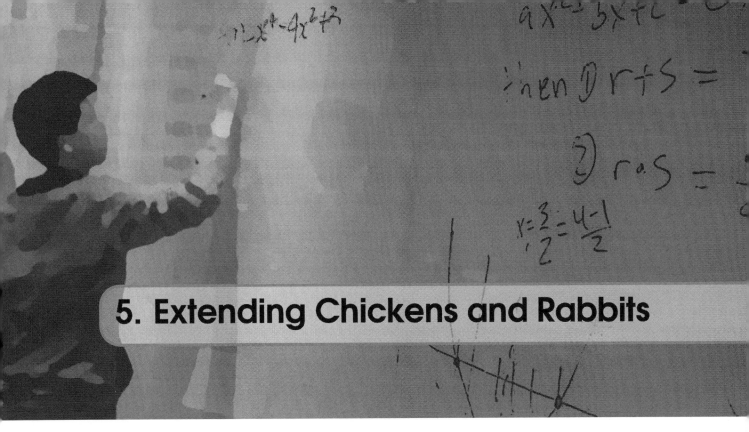

5. Extending Chickens and Rabbits

5.1 Example Questions

Problem 5.1 Review of the Chicken and Rabbit Method

(a) In Bob's Cycle Shop, workers received the delivery of an order Bob placed for the single seat bicycles and tricycles. There are a total of 90 seats and 215 wheels, plus all the other necessary parts and accessories. How many bicycles and how many tricycles can they assemble?

(b) There are 100 birds and cats. The birds have 80 more legs than the cats. How many birds and how many cats are there?

Problem 5.2 Applications to Mixing Problems

(a) Connor is helping his biology teacher get ready for the class's next laboratory. The teacher needs 10 liters of a 68% alcohol solution for the students to disinfect their tools. Connor makes this solution by mixing a 80% alcohol solution with a 50% alcohol solution. How many liters of each solution did Connor use?

(b) Debbie is mixing orange juice concentrate for her restaurant. The first juice concentrate is 64% real orange juice. The second is only 48% real orange juice. How many ounces of 48% real orange juice should she use to make 1600 ounces of 58% real juice?

Problem 5.3 Answer the following

(a) The owner of the Fancy Food Shoppe wishes to mix cashews selling at $8.00 per kilogram and pecans selling at $7.00 per kilogram. How much of each kind of nut should be mixed to get 8 kg worth $7.25 per kilogram?

(b) A meat distributor paid $2.50 per pound for hamburger meat and $4.50 per pound for ground sirloin. How many pounds of each did he use to make 100 pounds of meat mixture that will cost $3.24 per pound?

Problem 5.4 Solve the following questions.

(a) A candy shop sold three flavors of candies, cherry, strawberry, and watermelon, in the morning. The prices are $20/kg, $25/kg, and $30/kg, respectively. The shop sold a total of 100 kg and received $2570. It is known that the total sale of cherry and watermelon flavor candies combined is $1970. How many kilograms of watermelon flavor candies were sold?

(b) The school purchases 3 different sizes of projectors, total of 47. The large size costs $700, the medium costs $300, and the small costs $200. The total cost of the projectors is $21200, and there are twice as many medium projectors than the small. How many large projectors does the school purchase?

Problem 5.5 Connor is classifying bugs for a biology project. He has 15 bugs in total. Some are spiders, with 8 legs. Some are houseflies, with 6 legs and 1 pair of wings. The remaining are dragonflies, with 6 legs and 2 pairs of wings. If there are 98 legs and 17 pairs of wings, how many of each type of bug does Connor have?

Problem 5.6 A crab has 10 legs. A mantis has 6 legs and 1 pair of wings. A dragonfly has 6 legs and 2 pairs of wings. There are a total of 37 of the three types. There are 250 legs in total. There are 52 pairs of wings in total. How many of each kind are there?

Problem 5.7 A turtle has 4 legs and a crane has 2 legs. There are totally 100 heads of turtles and cranes, and there are 20 more crane legs than turtle legs. How many of each animal are there?

Problem 5.8 A party store has 100 balloons coming in either large or small sizes. The large balloons have a volume of 5 liters and are filled with 80% helium and 20% regular air. The small balloons have a volume of 2 liters and are filled with 90% helium and 10% regular air. If 312 liters of pure helium was used to fill the balloons, how many liters of regular air were used to fill the balloons?

Problem 5.9 Some chickens and rabbits have a total of 100 feet. If each chicken was exchanged for a rabbit, and each rabbit was exchanged for a chicken, there would be a total of 86 feet. How many chickens are there? How many rabbits?

Problem 5.10 Bella goes shopping at the marketplace for shawls and belts. The shawls she likes each cost $12. The belts she likes each cost $14. Bella has exactly enough money to buy a certain number of shawls. If she buys belts instead, she has exactly enough money to buy 3 fewer belts. How much money did Bella bring with her to the market?

5.2 **Quick Response Questions**

Problem 5.11 In a farm there are goats and ducks. The total number of heads is 100, and the total number of legs is 316. How many ducks are there?

Problem 5.12 Each set of chess is played by 2 students, and each set of Chinese checkers is played by 6 students. A total of 26 sets of chess and Chinese checkers are played by exactly 120 students in a school event. How many chess sets are there?

Problem 5.13 Tiffany scored 29 points in her school's playoff basketball game. She made a combination of 2-point shots and 3-point shots during the game. If she made a total of 11 shots, how many 3-point shots did she make?

Problem 5.14 The Math Zoom Camp allows students to form four-person teams or six-person teams to work together on projects. If there are 42 teams formed with the 200 total students in the camp, how many four-person teams are formed?

Problem 5.15 There are 48 tables in a restaurant. Small tables can seat 2 people, and big tables can seat 5 people. They can accommodate a maximum number of 159 people. There are more small tables than large tables at the restaurant. How many more are there?

Problem 5.16 17 people went to a farm. There were goats to feed and chickens to feed, and each person fed exactly one animal. It costs $1.50 to feed a chicken and $2.00 to feed a goat. In total, the people spent $32.50. How many chickens did they feed?

Problem 5.17 A convenience store owner wishes to mix together raisins and roasted peanuts to produce a high energy snack for hikers. The raisins sell for $3.50 per kilogram and the nuts sell for $4.75 per kilogram. Raisins and peanuts are mixed together mixed together to obtain 20 kg of this snack with a price of $4.00 per kilogram. How many pounds of peanuts were used?

Problem 5.18 Hank has a bottle of diluted syrup that is 60% maple syrup and a bottle of pure syrup that is 100% maple syrup in his restaurant. How many ounces of pure syrup should he mix with the diluted syrup in order to make 100 ounces of 85% maple syrup? Express your answer as a decimal rounded to the nearest hundredth if necessary.

Problem 5.19 How many gallons of 60% antifreeze should be mixed with 40% antifreeze to make 80 gallons of 45% antifreeze?

Problem 5.20 Cindy collects 20 insects for her biology class, all of which are spiders, dragonflies, and cicadas. (Note that a spider has 8 legs and no wings, a dragonfly has 6 legs and 4 wings, and a cicada has 6 legs and 2 wings.) She counts 138 legs and 36 wings altogether. How many cicadas does Cindy collect in total?

5.3 Practice Questions

Problem 5.21 On a good day, Chris the Squirrel picks 20 hazelnuts. On a rainy day he only picks 12 hazelnuts. During a few consecutive days he picked a total of 120 hazelnuts with an average of 15 per day. How many days were rainy?

Problem 5.22 How many gallons of a 25% alcohol solution must be mixed with a 50% alcohol solution to make 30 gallons of a 40% alcohol solution?

Problem 5.23 A butcher has some hamburger meat that is 4% fat and some hamburger meat that is 20% fat. How much of each type will he need to make 120 pounds of hamburger meat which is 10% fat?

Problem 5.24 Lily spent $490 to buy 80 color pencils for her art class, including red, green. and blue colors. The red pencils cost $2 each, the green ones cost $5 each, and the blue ones cost $10 each. Suppose she bought the same number of green and blue pencils. How many of each type of pencils did she buy?

Problem 5.25 A spider has 8 legs. A firefly has 6 legs and 2 pairs of wings. A cicada has 6 legs and 1 pair of wings. There are a total of 16 bugs of the three types. There are 110 legs in total. There are 14 pairs of wings in total. How many of each kind of bug are there?

Problem 5.26 A spider has 8 legs and no wings. A dragonfly has 6 legs and 2 pairs of wings. A cicada has 6 legs and one pair of wings. There are a total of 18 bugs of these types, with 118 legs and 20 pairs of wings. How many dragonflies are there?

Problem 5.27 The capacity of a big container is 4 gallons, and that of a small container is 2 gallons. 50 containers are filled with water, and there are totally 20 more gallons of water in the big containers than the small containers. How many big and small containers are there respectively?

Problem 5.28 A large bottle can hold 4 liters of oil, while every two small bottle can hold 1 liter of oil. A store has 100 liters of oil and the oil exactly fills up 60 bottles. How many of each kind of bottle does the store have?

Problem 5.29 Mike's cycle shop sells both bicycles (with two wheels) and tricycles (with three wheels). On Friday, Mike sold some bicycles and tricycles which combined have a total of 85 wheels. All the customers where very happy with their purchase, and on Monday, everyone who bought a bicycle on Friday came to buy a tricycle and everyone who bought a tricycle came to buy a bicycle. If these cycles sold on Monday had a total of 90 wheels, how many bicycles were bought on Friday?

Problem 5.30 Farmer George needs to buy some new roosters and hens for his farm. Each rooster costs $30 and each hen costs $42. George has enough money to buy some number of hens, but if he buys roosters instead he can buy 4 extra roosters. How much money does Farmer George have?

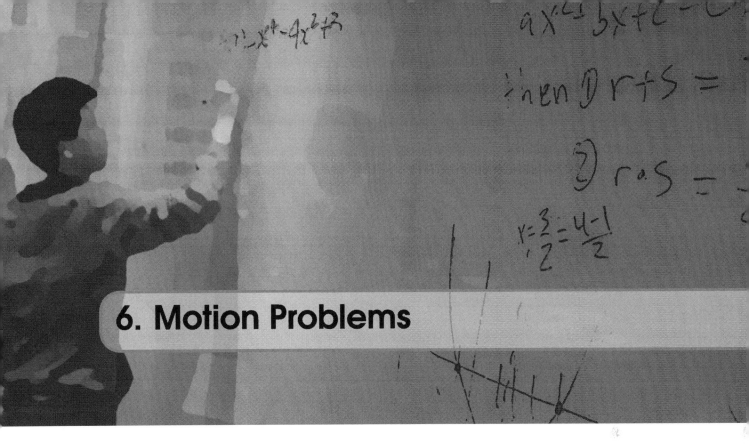

6. Motion Problems

6.1 Example Questions

Problem 6.1 Introductory Motion Questions

(a) Samantha needs to run a mile for less than 10 minutes for her PE class. What is the minimum average speed she must run to complete the mile in time? Give your answer in both miles per hour and meters per second (use the approximation 1 mile is roughly 1600 meters).

(b) Jimmy and his brother took a circular ride at an amusement park on averages of 30 miles per hour and took them $2\frac{1}{2}$ minutes. Roughly how big is the diameter of the circular track?

Problem 6.2 The Winchers family is taking a road trip from Los Angeles, California to Phoenix, Arizona. The distance is about 360 miles.

(a) If they drove at a constant speed of 60 miles per hour for the first 150 miles, and drove at a constant speed of 70 miles per hour for the remaining 210 miles, how long did it take them to get there?

(b) Suppose instead the family plans to get there in 6 hours. If they drove at a constant speed of 50 miles per hour for the first 120 miles, at what constant speed do they need to drive for the remaining 240 miles in order to get there in time?

Problem 6.3 Relative Speeds

(a) John's house and Mary's house are 14 miles apart. They start at noon to walk toward each other in order to go to a book fair together. John walks at a rate of 3 mph, and Mary walks at a rate of 4 mph. How many hours will it take them to meet?

(b) Terry and Susan are entered in a 24-mile race. Susan's average rate is 4 miles per hour and Terry's average rate is 6 miles per hour. Both start at the same time. How far will Susan be away from the finish line when Terry crosses the line?

Problem 6.4 Round Trips

(a) Mr. Winchers drives a car from home to a client site in Wilmington early morning at the speed of 72 miles per hour. On his way back home from the client site in Wilmington, the traffic is getting bad; he drives the car back home at a reduced speed of 48 miles per hour. What is his average speed for the round trip?

(b) Mike drives his car for a round trip between LA and San Diego. He drives at 70 miles per hour to get from LA to San Diego. At what speed should he drive back, if his average speed for the round trip is 60 miles per hour?

Problem 6.5 One day, Bob rode his bike to school. When school is off, he forgot his bike and walked home instead. He spent a total of 50 minutes on the road for the round trip. If he walked for both directions, he would have spent a total of 70 minutes. How much would be the total time if he rode his bike for both directions?

Problem 6.6 Ling goes mountain hiking in a park. She first walks uphill at a speed of 2.5 miles per hour, and she next walks downhill at a speed of 4 miles per hour. The round trip takes 3.9 hours. What is the distance for the round trip?

Problem 6.7 At 6 AM, bus station A starts to dispatch buses to station B, and station B starts to dispatch buses to station A. They each dispatch one bus to the other station every 8 minutes. The one-way trip takes 45 minutes. One passenger gets on the bus at station A at 6:16 AM. How many buses coming from station B will the passenger see en route?

Problem 6.8 Non-Constant Motion

(a) Sam walks up a hill. After every 30 minutes of walking he takes 10 minutes to rest. When he walks down the hill, he instead rests for 5 minutes after every 30 minutes of walking. Sam walks downhill 1.5 times faster than he walks uphill. If he spends 3 hours and 50 minutes traveling up the hill, how much time does he spend traveling down the hill?

(b) A hunting dog chases a hare 21 meters ahead. The dog runs in a series of jumps, with each jump being 3 meters long. Each jump for the hare is 2.1 meters. If the dog jumps three times for every four times the hare jumps, how much farther can the hare travel before the dog catches it?

Problem 6.9 Answer the following questions.

(a) Cindy rides her bike from home to school at a speed that is 120 meters per minute faster than if she walks, and the time she spends is 3/5 less than if she walks. How fast does Cindy walk from home to school?

(b) Joe and JoAnn walk toward each other from two locations that are 36 miles apart. If Joe departed 2 hours earlier, they would meet 2.5 hours after JoAnn departed. If JoAnn departed 2 hours earlier, they would meet 3 hours after Joe departed. Find the respective speed at which each walks.

Problem 6.10 Two trains are 121 meters and 99 meters in length respectively. They travel at a rate of 40 km/h and the other at the rate of 32 km/h.

(a) If the trains are moving in opposite directions, how long would it take them to completely clear each other from the moment they meet?

(b) If the two trains are traveling in the same direction, how long would it take them to completely clear each other, if the faster train has just met up with the back of the slower train?

6.2 Quick Response Questions

Problem 6.11 Bob drives for 350 miles going 50 mph and then another 350 miles going 70 mph. Is it true that his average speed for the full trip is 60 mph?

Problem 6.12 Bob drives for 6 hours going 50 mph and then another 6 hours going 70 mph. Is it true that his average speed for the full trip is 60 mph?

Problem 6.13 Larry and Moe go running. The time Larry runs is twice that of Moe, but Moe runs 3 times faster. Who runs farther and why?

(A) They run the same distance.
(B) Larry because he runs for more time than Moe.
(C) Moe because he runs faster than Larry.
(D) Moe because he runs 3 times faster than Larry who only runs twice the time.

Problem 6.14 A bat is flying at a speed of 45 kilometers per hour. How much time does it take to travel a distance of 1,800 kilometers? Give your answer in hours.

Problem 6.15 Adam takes a train to go visit a friend who lives in a city that is 360 kilometers away. The train left his home station at 8:35AM, and arrived at the destination station at 1:05PM. How fast has the train traveled measured by average speed in km/hr?

Problem 6.16 Joe and his family are planning to go to a national park which is 600 miles away from the home. How fast in miles per hour must they drive if they want to get there in 15 hours?

Problem 6.17 Jasmine took a walk after dinner. She first walked 5 km in 1.5 hours, and then walked for 1 km in 0.5 hour in the same direction. What is her average speed for the whole journey in km/hr?

Problem 6.18 Suppose a truck travels in segments that are described in the table below:

Segment	Distance (miles)	Time (hours)
1	30	1
2	90	2
3	50	1

What is the average speed of the truck in miles per hour?

Problem 6.19 Two friends leave the same place at the same time traveling in the same direction. One travels at a speed of 55 mph and the other travels at a rate of 65 mph. After 2 hours, how many miles will they be away from each other?

Problem 6.20 Katie went hiking on a hill near her home. From the bottom of the hill, She went up to the top and then came down along the same trail, back to the spot she started. Assume her uphill speed was 3 miles per hour, and her downhill speed was 6 miles per hour. What is her average speed in mph for the whole uphill-downhill trip?

6.3 Practice Questions

Problem 6.21 A boat has a rip-hole in the bottom while 20 miles away from the shore. The water comes in at a rate of 1.5 tons every minute, and the boat would sink after 70 tons of water came in. How fast must the boat go in order to reach the shore before sinking?

Problem 6.22 Frank drives his car for a distance of 300 miles. For the first 135 miles, he drives at a constant speed of 45 miles per hour. At what constant speed does he drive for the remaining distance to average 50 miles per hour?

Problem 6.23 Tyler and Hannah start to walk on the same direction from the same place. Tyler walks at 5 miles per hour. Hannah walks at 1 mile per hour for the first hour, 2 miles per hour for the second hour. Hannah increases her speed by 1 mile per hour after each hour. How long does it take for Hannah to catch up with Tyler?

Problem 6.24 Thomas and his family went on a road trip last week. They traveled 50 mph from Chicago, IL to Minneapolis, MN and 65 mph on the return trip. What was the average speed for the entire round trip?

Problem 6.25 It takes 40 minutes for Dave to walk from home to school. It takes 15 minutes if he rides a bike instead. One day, he first rides a bike for 9 minutes before the bike breaks. He then walks the remaining distance to school. How much total time does Dave spend getting to school?

Problem 6.26 If I drive from Irvine to Fullerton at 60 miles per hour and then from Fullerton to Irvine at 40 miles per hour, what is my average speed for the whole journey?

Problem 6.27 Stephanie begins walking at a pace of 4 km per hour from one end of the trail that is 34 km long. Bob begins at the other end of the trail at the same time, walking towards Stephanie at a pace of 6 km. How long will it take for them to pass each other?

Problem 6.28 Omar walks up a hill. After every 40 minutes of walking uphill he takes 10 minutes to rest. Downhill he rests for 5 minutes after every 40 minutes of walking. Omar walks downhill at a speed 2 times as fast as that he walks uphill. If he spends 2 hours traveling down the hill, how much time does he spend traveling up the hill?

Problem 6.29 Sami and Rajan practice running together. If Sami starts to run after Rajan runs for 10 meters, then it will take Sami 5 seconds to catch up with Rajan. If Sami starts to run after Rajan runs for 2 seconds, then it will take Sami 4 seconds to catch up with Rajan. How fast can each person run?

Problem 6.30 A train, 110 meters in length, travels at 60km/h. In what time will it pass a man who is walking at 6km/h (i) against it; (ii) in the same direction?

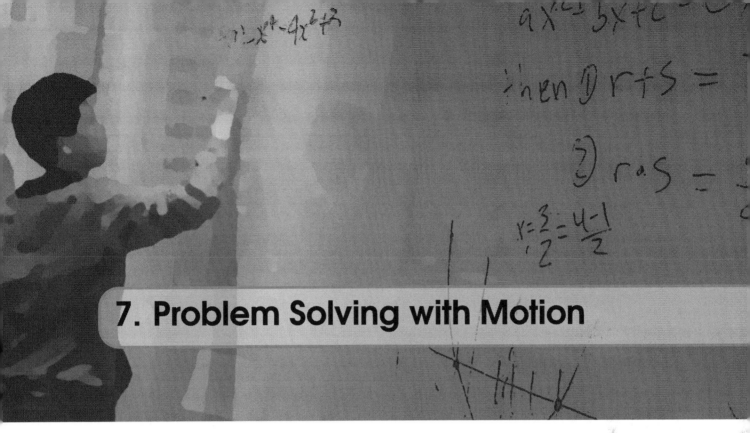

7. Problem Solving with Motion

7.1 Example Questions

Problem 7.1 Motion Warmups and Review

(a) A tennis ball is thrown a distance of 20 meters. What is its speed if it takes 0.5 seconds to cover the distance?

(b) It takes 6 hours for an airplane to fly a round trip. If the speed of the airplane is 1500 km per hour on the departure trip, and 1200 km per hour on the return trip. What is the one-way distance ?

Problem 7.2 Connor and his friend Derek went to river tubing at the Math Zoom Summer Camp. Connor sat still on the tube and enjoyed the beautiful scenes on the river bank, with no effort made to pilot the tube. Derek, however, paddled down the river. Both left the starting point at 3 pm and traveled a total distance of two and a quarter miles.

(a) Connor arrived at 4:15 pm, how fast was the water flowing in the river? Use miles per hour for the speed.

(b) Derek ended up arriving at the ending point at 3:45 pm. How fast can Derek paddle in still water?

Problem 7.3 Answer the following.

(a) It takes a ship 6 hours to travel downstream between two piers, and 8 hours upstream. If the water flows at a speed of 2.5 miles per hour, at what speed would the ship travel in still water?

(b) Mr. and Mrs. Winchers were rowing a boat upstream in the river. At exactly 1 pm, Mrs. Winchers's hat fell into the water, but they didn't find out until sometime later. They immediately turned around and rowed downstream, and caught up with the hat 6 miles from the point the hat fell. Assuming the water flow speed is 4 miles per hour. (1) What is the exact time when they found the hat? (2) Assuming they rowed at a constant speed relative to the water, what was the time when they realized the hat fell and turned around?

Problem 7.4 Linda goes mountain biking in a park. She first bikes on flat road at a speed of 12 miles per hour, then goes uphill at a speed of 9 miles per hour, and she next bikes downhill at a speed of 18 miles per hour, along the same trail as the uphill trip. Finally she goes back home along the same flat road she traveled earlier. The round trip takes 4 hours. What's the distance for the round trip?

Problem 7.5 Starting at the same time, Heather and Brenda drive their cars from site A toward site B. Heather drives at 52 km per hour, and Brenda drives at 40 km per hour. After 6 hours of driving, Heather passes a truck traveling in the opposite direction. One hour later, Brenda passes the same truck still traveling in the opposite direction. At what speed does the truck travel?

Problem 7.6 Brian and David run along a circular track, starting from the same point, going opposite directions. They meet after 36 seconds. Assume that David runs the whole circle in 90 seconds. How long does it take Brian to run the whole circle?

Problem 7.7 Starting at the same time, Cathy and David drive two cars toward each other from the two ends, call them A and B, of the same road. Cathy drives 1.2 times faster than David. When they pass by each other, they are 8 miles away from the halfway point between A and B. Find the total length of the road.

Problem 7.8 A railroad bridge measures 1000 meters long. A train passes the bridge. It takes 120 seconds from the time the train enters the bridge to the time the whole train gets off the bridge. There are 80 seconds during which time the whole train is on the bridge. Find both the speed and the length of the train.

Problem 7.9 A road consists of uphill, flat and downhill sections with that order. The distances of the three sections are in ratio 1 : 2 : 3 with a total distance of 20 miles. The times JoAnn spends on the three sections are in ratio 4 : 5 : 6. She walks at a speed of 2.5 miles per hour uphill. What is the total time she spends on the road?

Problem 7.10 Alice, Bob, and Cindy drive their cars separately from site A to site B simultaneously. Alice drives at 60 mph and Bob drives at 48 mph. Alice passes a car from the opposite direction after 6 hours of driving. One hour later, Bob pass the same car still traveling in the opposite direction. One more hour later, Cindy also passes the same car. Find the speed at which Cindy drives her car.

7.2 **Quick Response Questions**

Problem 7.11 Suppose a train travels a distance of 120 miles in 3 hours. What is the average speed of the train in miles per hour?

Problem 7.12 Mary and Amy rollerblade at an average speed of 9 miles per hour for 3.5 hours, how many miles will they travel?

Problem 7.13 Sam begins walking at a pace of 4 km per hour from one end of a trail that is 34 km long. Ashley begins one hour later at the other end of the trail, walking towards Sam at a pace of 6 km per hour. How long will it take for them to pass each other in hours?

Problem 7.14 Tom drives his car for a round trip between place A and place B. He drives at 40 km per hour to get from A to B. At what speed should he drive back from B to A, if his average speed for the round trip is 48 km per hour? Give your answer in km/h.

Problem 7.15 Water flows in a river at 3 m/s. Traveling upstream, George's speed boat is able to travel 1.8 km in 3 minutes. How many m/s can George's boat travel in still water?

Problem 7.16 In a cross-country race, Tony drove his car for 707 kilometers in 7 hours. What was his average speed in kilometers per hour?

Problem 7.17 David and Ray hike a mountain trail in Crystal Cove. David starts out on the trail at a pace of 4 kilometers per hour. One hour later, Ray starts out on the same trail at 6 kilometer per hour. How many hours will it take Ray to catch up to David?

Problem 7.18 It takes a pigeon 2 hours to fly with the wind between two houses, and 3 hours against the wind. If the wind blows at a speed of 2.5 miles per hour, at what speed in mph would the pigeon travel in a windless day?

Problem 7.19 A bridge consists of three sections of equal length: an uphill section, a flat section and a downhill section. At what average speed does a cyclist ride his bicycle if he travels the three sections at a speed of 4 meters per second, 6 meters per second and 8 meters per second respectively? Round your answer to the nearest whole number.

Problem 7.20 A train moving 25 km/h takes 18 seconds to pass a platform. Next, it takes 12 seconds to pass a man walking at 5 km/h in the opposite direction. Find the length of the train in meters.

7.3 Practice Questions

Problem 7.21 Melisa drove for 3 hours at a rate of 50 miles per hour and for 2 hours at 60 miles per hour. What was her average speed for the whole journey?

Problem 7.22 A boat takes 3 days to travel from town A to town B, but it takes 4 days to travel from town B to town A. If a motor-less raft is left alone in the water by town A, how long will it take for the raft to float to town B?

Problem 7.23 Adam lives in Town A, and Bob lives in Town B, and the two towns are 10 miles apart. At the same time, Adam starts walking from Town A to Town B at 3 miles per hour, and Bob starts walking from Town B to Town A at 2 miles per hour, and a bird also starts flying from Adam towards Bob. Once the bird meets Bob, it turns back towards Adam, and once it meets Adam it turns towards Bob again, and so on, until Adam and Bob meet. Assuming the bird flies at 15 miles per hour. What is the total distance that the bird flies?

Problem 7.24 An ant crawls along the sides of an equilateral triangle. It starts at one vertex and crawls at a rate per minute of 50 cm, 20 cm and 40 cm, respectively, on each of the three sides of the triangle. What is the ant's average speed as it travels around the triangle?

Problem 7.25 As Emily is riding her bicycle on a long straight road, she spots Emerson skating in the same direction 1/2 mile in front of her. After she passes him, she can see him in her rear view mirror until he is 1/2 mile behind her. Emily rides at a constant rate of 12 miles per hour, and Emerson skates at a constant rate of 8 miles per hour. For how many minutes can Emily see Emerson?

Problem 7.26 Alice leaves site A toward site B at the same time Bob leaves site B toward A. Alice drives at 40 miles per hour, and Bob drives at 60 miles per hour. After they pass by each other, Alice drives 4.5 additional hours to arrive at B. How far is it between A and B?

Problem 7.27 Starting at the same time, Tom and Jerry walk toward each other. Tom walks from site A to B at 5 miles per hour. Jerry walks from B to A. After they meet, Jerry walks an additional 10 miles to arrive at A, and Tom spends additional 1.6 hours to walk and arrive at B. What is the rate at which Jerry walks?

Problem 7.28 It takes 25 seconds for a train to pass completely pass through a tunnel which measures 250 meters long. It takes 23 seconds for the train to completely pass through another tunnel which measures 210 meters long. How long does it take the train to pass an approaching train which is 320 meters long and at the speed of 18 m/s?

Problem 7.29 George goes to school by riding his bike to the bus station, taking the bus, and then walking to his classroom. The ratio of the three distances is 2:8:1. His biking speed is 10 mph. The bus travels at a speed of 50 mph. His walking speed is 2 mph. What is his average speed in his way to school?

Problem 7.30 Starting at the same time, a bus and a truck start traveling toward each other. After 18 hours the two vehicles meet. The bus travels at 50 miles per hour. The truck travels at 42 miles per hour, but stops for a 1 hour break after every 3 hours of travel. What is the distance between the two starting locations?

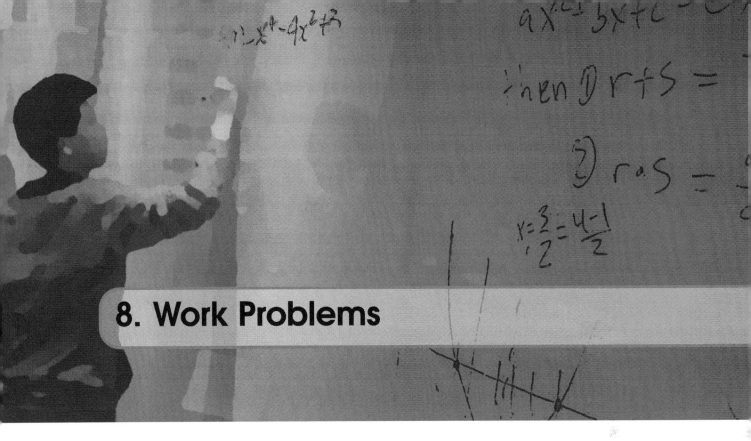

8. Work Problems

8.1 Example Questions

Problem 8.1 Introductory Work Questions

(a) Samantha invites Jo to help with her baking. Samantha usually prepares the pretzels for 40 minutes before putting them to the oven. Today she has Jo for help, so they worked together to prepare the pretzels. Being less experienced, Jo would have taken 60 minutes to prepare the pretzels all by herself. Working together, how long would it take for them to prepare the pretzels? In addition, it takes 15 minutes for the prepared pretzels to be baked. What is the total time needed for Samantha and Jo to prepare and bake the pretzels together?

(b) Suppose Chris can paint the entire house in fourteen hours, and Bill can do it in ten hours. How long would it take the two painters working together to paint the house?

Problem 8.2 Working Together, Working Alone

(a) Tom and Jerry can finish organizing the books at school's library together in 5 hours. If Tom do it alone, it will take him 8 hours. How long would it take Jerry to finish the same task alone?

(b) Stan can load the truck in 40 minutes. If I help him, it takes us 15 minutes. How long will it take me alone?

Problem 8.3 Patty and Tracy can finish decorating a house for the holidays in 2.5 hours if they work together. Patty works twice as fast as Tracy. How long would it take to each of them if they work alone?

Problem 8.4 Phil can paint the garage in 12 hours and Rick can do it in 10 hours. They work together for 3 hours. How long will it take Rick to finish the job alone?

Problem 8.5 Larger Groups

(a) James, Patty, and Joseph can organize the new products in the warehouse in 2 hours. If James does the job alone, he can finish in 5 hours. If Patty does the job alone, she can finish it in 6 hours. How long will it take for Joseph to finish the job alone?

(b) Mona can complete a task alone in 150 minutes. Sarah can finish the same task in 3 hours. They work together for 30 minutes, and then a new worker, Li, joins them, and they finish the task 30 minute later. How long would it take Li to finish the task alone?

Problem 8.6 Anthony can cut a lawn in 2 hours, Mia can cut the same lawn in 3 hours, and Dandria can cut the same lawn in 2 hours. Anthony cuts the lawn for $\frac{1}{2}$ hour, and then Mia replaces Anthony and cuts the lawn for 1 hour herself. How many additional minutes will it take Dandria to finish cutting the lawn by herself?

Problem 8.7 Suppose if Ethan works for 5 days and Owen for 6 days, they finish a project. Alternatively, if Ethan works for 7 days and Owen for 2 days, they also finish the project. How long does it take for Ethan alone to complete the project? For Owen alone?

Problem 8.8 Alternating Work

(a) It takes Elizabeth 9 hours to complete a project. For Tiffany, it takes 12 hours. If they take turns, starting with Elizabeth, each working for one hour at a time, how much total time does it take for them to complete the project?

(b) It takes Rianna 24 days to finish a job. For Helen, it takes 32 days. Rianna works on it for some days before Helen takes it over, and it takes a combined total of 26 days to get the job done. How many days does Rianna work on the job?

Problem 8.9 Niki always leaves her cell phone on. If her cell phone is on but she is not actually using it, the battery will last for 24 hours. If she is using it constantly, the battery will last for only 3 hours. Since the last recharge, her phone has been on a total of 10 hours, and during that time she has used it constantly for 60 minutes. If she doesn't talk any more (but leaves the phone on), how many more hours will the battery last?

Problem 8.10 Students from the Key Club at Whitman High School wash cars from the two parking lots A and B. There are four times as many cars in lot A than in lot B. First, they wash the cars in lot A for half a day. Next, half of students continue, and half of students start to wash cars in lot B. The work in lot B is done at the end of the day. Unfortunately, there are still cars unwashed in lot A. If all the students worked together to finish the cars in lot A, how long would it take?

8.2 Quick Response Questions

Problem 8.11 If 8 people can finish a task in thirteen days, how many days would it take to do the same job with 16 people? Express your answer as a decimal.

Problem 8.12 Chris works for 2 hours and washes 200 dishes. Mike works for 3 hours and washes 200 dishes. How many dishes can they wash in total if they work for 6 hours each?

Problem 8.13 Jodi can finish $\frac{2}{3}$ of her chores in one hour. How long in total does it take Jodi to do her chores? Give your answer as a decimal, rounded to the nearest hundredth if necessary.

Problem 8.14 Andy and Jim usually work together at the end of each day to file reports. It takes them 45 minutes to file the reports together. However, Andy is sick today so Jim has to do the reports by himself. If Andy and Jim work at the same rate, how long will it take Jim to do the reports alone? Give your answer in minutes.

Problem 8.15 Laura can finish her math homework in 75 minutes. Which of the following is NOT true?

(A) Laura can finish $\frac{1}{75}$ of her homework in one minute
(B) Laura can finish $\frac{1}{2}$ of her homework in 35 minutes
(C) Laura could finish two homeworks in 150 minutes
(D) Laura can finish $\frac{1}{3}$ of her homework in 25 minutes

Problem 8.16 Eldridge can split a cord of wood in 4 days and his father can do it in 3 days. How long would it take them (in days) if they worked together?

(A) $\dfrac{7}{12}$

(B) 7

(C) 2

(D) $\dfrac{12}{7}$

Problem 8.17 Using a riding lawn mower, Abby can mow the lawn in 120 minutes. Her sister Carla takes 3 hours using an older mower. How many minutes will it take them if they work together?

Problem 8.18 Joe's mom can clean the kitchen in 45 minutes. If Joe helps his mother, they can clean it in 30 minutes. How long would it take Joe to clean it by himself in minutes?

Problem 8.19 Peter can paint a wall in 40 minutes and John can paint the wall in 60 minutes. If they work together for 12 minutes, what percentage of the wall is left unpainted? Round your answer to the nearest percent.

Problem 8.20 Amy and her sister Clair's house has a 420 square foot lawn in the back. Amy can mow 120 square feet in 30 minutes. When Amy and Clair work together, they can finish the whole lawn in one hour. How many square feet per minute can Clair mow?

8.3 Practice Questions

Problem 8.21 It takes 1.5 hours for Jim to water the plants in a garden. Lily can water the same amount of plants in 2 hours. How long will it take Jim and Lily, work together to water the plants in the garden?

Problem 8.22 Employee A can complete a task in 3 hours. When working with Employee B, they can complete it in 2 hours. How long does it take for Employee B to finish the task if he/she works alone?

Problem 8.23 Melisa can finish a project in 2 hours, while Terry works 1.5 times faster than Melisa. How long would it take them to finish together?

Problem 8.24 It will take a Type A robot 6 min to weld a fender, but a Type B robot takes only $5\frac{1}{2}$ minutes. If the robots work together for 2 min, how long will it take the Type B robot to finish welding by itself? Express your answer as a mixed number.

Problem 8.25 Bob, John, and Calvin can paint a wall alone in 2 hours, 2.5 hours, and 1.5 hours, respectively. How long does it take if all three of them work together?

Problem 8.26 Bill usually takes 50 minutes to groom the horses. After working for 10 minutes, he was joined by Ann and they finished the grooming in 15 minutes. How long would it have taken Ann working alone?

Problem 8.27 If Iris spends 3 days and Olivia spends 5 days on a project, $\frac{1}{2}$ of the work can be completed. If instead Iris spends 5 days and Olivia spends 3 days on the project, $\frac{1}{3}$ of the work can be completed. How long does it take to complete the whole project if Iris and Olivia work together?

Problem 8.28 It takes Jacqueline 50 minutes to type a draft. For Virginia, it takes 30 minutes. Suppose after Jacqueline types for some time, Virginia types the rest of the draft, and it takes a combined total of 42 minutes. What fraction of the draft did Jacqueline type?

Problem 8.29 Barry has to clean his room. If he focuses and works hard, it takes him 30 minutes to clean his room. However, Barry is easily distracted and when he his distracted it takes him a total of 2 hours to clean his room. Barry started cleaning his room 30 minutes ago but got distracted after 10 minutes. If he now stays focused, how long would it take him to finish cleaning his room?

Problem 8.30 Adam and Bob are each assigned a task to paint a wall. The two walls are identical. At the beginning, Adam went to the wrong wall and painted 500 square feet on Bob's wall. At this moment Bob came and found Adam's mistake. Adam then returned to his own wall and Bob continued painting his wall. After a few days, Bob finished his task, but Adam is not yet done with his job. Now Bob decided to come and help Adam. Bob painted 1000 square feet on Adam's wall. Which person did more of the job?

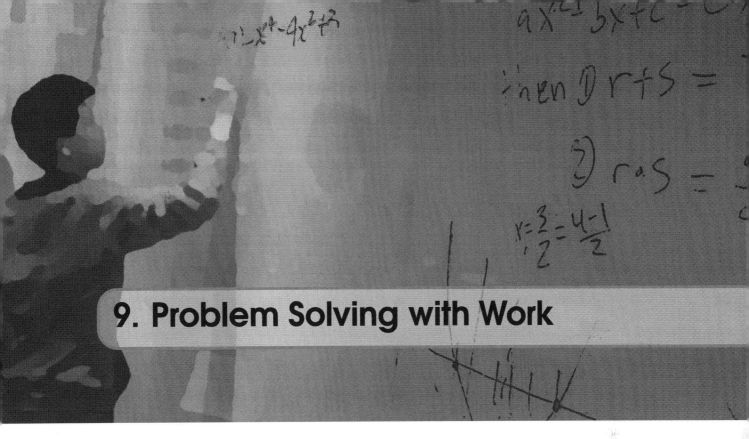

9. Problem Solving with Work

9.1 Example Questions

Problem 9.1 The community swimming pool held 18000 gallons of water when it was full. One day, Samantha went swimming but found that the pool was being drained for cleaning.

(a) If the draining pump drained the water at the rate of 12 gallons per minute, how long would it take for the pool to be completely drained?

(b) After the pool was drained, another water pump was used to fill the pool with fresh water. If the rate of filling the pool was 60 gallons per minute, how long would it take for the pool to be refilled with fresh water?

Problem 9.2 Filling and Draining Practice

(a) There is a 10,000 liter swimming pool in Lance's community. The pool has two pipes: A and B. Pipe B delivers 1,000 liters water per hour. When pipe A and B are both on, the pool can be filled in 4 hours. How many liters per hour can pipe A deliver?

(b) A pool has two inlet pumps A and B. If pump A alone is open, it takes 12 hours to fill the pool with water. If pump B is open, it takes 18 hours to fill the pool with water. If the pool needs to be filled in 10 hours, what is the least amount of time both pumps need to be open?

Problem 9.3 More Pumps and Drains

(a) There is a drain and a hose in the pool. It is known that the hose can fill the pool in 21 hours, and the drain can empty the pool in 24 hours. How long does it take to fill the pool if the drain is open at the same time?

(b) If pump A is used alone, it takes 6 hours to fill the pool. Pump B takes 8 hours alone to fill the same pool. Uncle Sam wants to use three pumps: A, B and C to fill the pool in 2 hours. What should be the rate of pump C in order to accomplish Uncle Sam's goal?

Problem 9.4 When two teams A and B work together, it takes 18 days to get a job completed. After team A works for 3 days, and team B works for 4 days, only $\frac{1}{5}$ of the job is done. How long does it take for team A alone to complete the job? For team B alone?

Problem 9.5 Brandon, Richard, and Samuel are friends. They build a wall divider in a yard. Brandon and Richard work together, and they build $\frac{1}{3}$ of the divider in 5 days. Next, Richard and Samuel work together, and they build $\frac{1}{4}$ of the rest of the divider in 2 days. Last, Brandon and Samuel work together, they finish the rest in 5 days. How long does it take for Brandon alone to build the divider? For Richard alone? For Samuel alone?

Problem 9.6 Calvin and Tony worked on producing a set of machines. Calvin planned to complete $\frac{7}{12}$ of the task. After he finished, he helped Tony to produce 24 pieces. The ratio of the number of pieces of Calvin to Tony is 5:3. How many pieces did Tony make?

Problem 9.7 Ratios with Work

(a) It takes $\frac{1}{3}$ more time for Andy to plant one tree than for Nathan. If Andy and Nathan work together, then in the end Nathan plants 36 more trees than Andy does. How many trees are in total?

(b) Carolyn reads a book. Initially, the number of pages that she had read and the number of pages that she has not read are in ratio $3 : 4$. After she reads an additional 33 pages of the book, the ratio becomes $5 : 3$. How many pages does the whole book have?

Problem 9.8 Parts and Pieces

(a) Robot A can produce 48 pieces of machinery per hour, and robot B can produce 36 pieces per hour. After they worked together for 8 hours, there were 64 pieces marked as defective when the company tested them. How many non-defective pieces they can produce together in one hour?

(b) A certain number of small parts need to be produced. 30 parts are scheduled to be produced after each day. After $\frac{1}{3}$ of the parts are produced, the rate of production increases by 10% thanks to improvement in efficiency. It takes 4 fewer days to produce all the parts than scheduled. How many parts are in total?

Problem 9.9 A project is planned to be completed by 45 people, and it will take some days to do it. After 6 days of work, 9 people left the team. As a result, it takes 4 more days to complete the project than originally planned. In how many days did they originally plan to finish the project?

Problem 9.10 There are two pumps A and B. They are used to fill water in two pools that, if full, hold equal amount of water. The ratio of the water-pumping rate for pump A and pump B is $7:5$. After $2\frac{1}{3}$ hours, the water in the two pools, if combined, can exactly fill one pool. Next pump A increases the water-pumping rate by 25%, pump B reduces the water-pumping rate by 30%. After pump A fills the pool, how much longer does it take for pump B to fill the second pool? Express your answer as a mixed number.

9.2 Quick Response Questions

Problem 9.11 Suppose George fills his bathtub with the faucet in 3 minutes. If his bathtub has a capacity of 75 liters, what is the rate (in liters per minute) of water leaving the faucet?

Problem 9.12 One drain pipe can empty a swimming pool in 6 hours. Another pipe takes 3 hours. If both pipes are used simultaneously to drain the pool, how long does it take the drain the pool? Give your answer in hours, rounded to the nearest hour.

Problem 9.13 Kathy takes 3 hours to wash 300 dishes, and Andrew takes 2.5 hours to wash 300 dishes. How many hours will it take to wash 1100 dishes if they work together?

Problem 9.14 Nancy can row a boat across the river in 45 minutes, while Susan can do it in 35. If both of them sit in one boat and row together, how long will it take?

(A) $\frac{1}{3}$ of an hour

(B) $\frac{16}{315}$ minutes

(C) 80 minutes

(D) $\frac{315}{16}$ minutes

Problem 9.15 Adam can mow his entire yard in three hours. His sister, Brooke, can mow $\frac{3}{4}$ of the same yard in two hours. Using two identical mowers, what part of the yard can they mow in one hour working together?

(A) $\dfrac{17}{24}$

(B) $\dfrac{5}{6}$

(C) $\dfrac{17}{12}$

(D) $\dfrac{1}{12}$

Problem 9.16 Lincoln can do a job in 8 hours and Dave can do it in 6 hours. What part of the job can they do by working together for 2 hours?

(A) $\dfrac{7}{24}$

(B) $\dfrac{5}{6}$

(C) $\dfrac{7}{12}$

(D) $\dfrac{1}{14}$

Problem 9.17 Billy and Tim can paint a fence in 4 hours together. It is known that Billy can paint the same fence alone in 6 hours. How long would it take Tim to paint the fence alone? Round your answer to the nearest hour if necessary.

Problem 9.18 Two mechanics in the maintenance department were working on Daniel's car. One can complete the maintenance service in 4 hours, but the other mechanic, who is new, needs 8 hours. How long would it take the two mechanics working together to finish the service? Give your answer in minutes. Round to the closest minute if necessary.

Problem 9.19 It takes pump A 6 hours to fill the pool, pump B 8 hours, and pump C 4.8 hours. How long (in hours) would it take the three pumps together to fill the pool? Round your answer to the nearest tenth.

Problem 9.20 Janelle cleans her aquarium by replacing $\frac{2}{3}$ of the water with new water, but that doesn't clean the aquarium to her satisfaction. She decides to repeat the process, again replacing $\frac{2}{3}$ of the water with new water. How many times will Janelle have to do this so that at least 95% of the water is new water?

9.3 Practice Questions

Problem 9.21 The community swimming pool holds 18000 gallons of water when it was full. It can be drained at 12 gallons per minute and filled at 60 gallons per minute. Samantha had a strange thought: what if while refilling the empty pool, the people forgot to turn off the draining pump while pumping in fresh water? How long would it take to refill the pool?

Problem 9.22 One pipe can fill a swimming pool 1.5 times faster than a second pipe. If the gardener opens both pipes, they fill the pool in 5 hours. How long would it take to fill the pool if only the slower pipe is used? How about only the faster pipe?

Problem 9.23 A pool can be filled by pipe A in 3 hours, and pipe B in 5 hours. When the pool is full, it can be drained by pipe C in 4 hours. Suppose the pool is empty and all three pipes are open, how long will it take to fill up the pool?

Problem 9.24 Ben and Jack can finish a task in 6 days working together. If Ben works alone for 5 days and then Jack takes over and works for 3 days, they finish $\frac{7}{10}$ of the work. How long would it take for each of them complete the task working alone?

Problem 9.25 If Emily and Julia work together, they can finish a project in 6 days. It takes same amount of time for Emily to complete $\frac{1}{2}$ of the project as it takes for Julia to complete $\frac{1}{3}$ of the project. How long does it take for Emily alone to complete the project?

Problem 9.26 Peter painted $\frac{1}{3}$ of a room while Richard painted $\frac{2}{5}$ of the same room. It then took Peter 1 hour, 40 minutes to finish painting the remainder of the room by himself. In how many hours could Peter have painted the entire room by himself? Express your answer as a mixed number.

Problem 9.27 A senior worker and a new worker together produce a set of machines. The senior worker can produce 40 pieces per hour, and the new worker can produce 30 pieces per hour. When they finished, the new worker produced exactly 450 pieces. How many pieces did they produce in total?

Problem 9.28 A senior worker and a new worker worked together for 2 days and finished $\frac{3}{5}$ of their work. The senior worker then took 2 days off while the new worker continued. After that, the senior worker went back the two workers worked together to finish their work. If the senior worker works twice as fast as the new worker, how many days did it take in total to finish the work? Express your answer as a mixed number.

Problem 9.29 Company A and B plan to manufacture a number of TVs together. After Company A worked for 6 days, it had finished $\frac{1}{4}$ of the TVs. Then the two companies worked together and finished the rest of the TVs in 6 days. It is known that Company B can produce 80 TVs per day. How many TVs in total did the two companies produce?

Problem 9.30 A pool has an inlet pump and an outlet pump. If the pool is empty, and the inlet pump is open, it takes 5 hours to fill the pool with water. If the pool is full, and the outlet pump is opened, it takes 7 hours to empty the pool. Suppose after the inlet pump is open for 2 hours, both the pumps are opened. How much longer does it take for the pool to be half full of water?

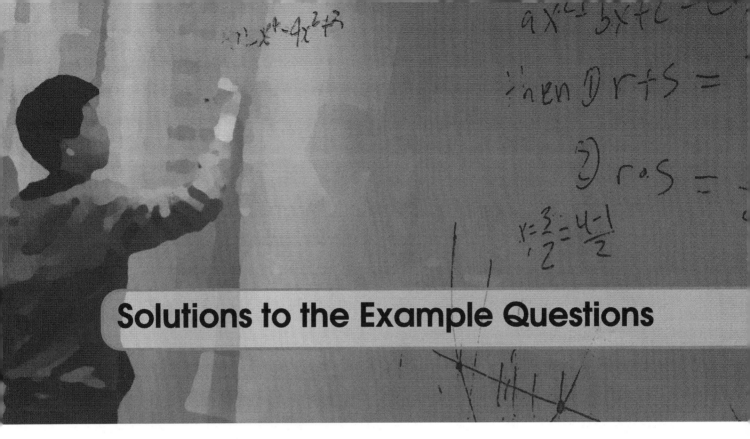

Solutions to the Example Questions

In the sections below you will find solutions to all of the Example Questions contained in this book.

Quick Response and Practice questions are meant to be used for homework, so their answers and solutions are not included. Teachers or math coaches may contact Areteem at info@areteem.org for answer keys and options for purchasing a Teachers' Edition of the course.

1 Solutions to Chapter 1 Examples

Problem 1.1 Find x in the following proportions.

(a) $12 : 17 = x : 68$.

Answer

48

Solution

Note that $68 = 17 \times 4$, so to make the ratios equal we must have $x = 12 \times 4 = 48$.

(b) $1.68 : 12 = 7 : x$.

Answer

50

Solution

Working directly, note that because $168 = 24 \times 7$, we have that $1.68 \div 7 = 0.24$. Therefore

$$x = 12 \div 0.24 = 50.$$

(c) $95 : 6.5 = x : 5.2$

Answer

76

Solution

First note that if $95 : 6.5 = x : 5.2$ we also know that $95 : 65 = x : 52$ if we multiply the second numbers in each ratio by 10. Simplifying $95 : 65 = 19 : 13$ (dividing by 5). Hence, as $13 \times 4 = 52$, we see that $x = 19 \times 4 = 76$.

Problem 1.2 Solve the following involving ratios.

(a) Thomas mixed 3 pints of red paint with 4 pints of blue paint to make a new color.

He now wants to use 27 pints of red paint and some blue paint to make the same color. How many pints of blue paint will he need?

Answer

36

Solution 1

We know the ratio of red to blue paint is $3:4$. Since

$$27 \div 3 = 9$$

we can multiply both sides of the ratio by 9 to get

$$3:4 = 27:36.$$

Therefore, if Thomas uses 27 pints of red paint he will need 36 pints of blue paint.

Solution 2

Let the amount of blue paint Thomas needs be x pints. Since the ratio of red to blue paint is $3:4$ and Thomas wants to use 27 pints of red paint we have

$$\frac{3}{4} = \frac{27}{x}.$$

Cross multiplying we get

$$3x = 108$$

so solving for x we have

$$x = 36.$$

so Thomas will need 36 pints of blue paint.

(b) The ratio of boys to girls in Math Zoom Academy is $5:3$. If there are 24 more boys than girls in Math Zoom Academy, then how many girls are in Math Zoom Academy?

Answer

36

Solution 1

The ratio of boys to girls is $5:3$, so there are 2 more boys than girls for every 3 girls. Since

$$24 \div 2 = 12$$

multiplying both sides of the ratio by 12 we get

$$5:3 = 60:36$$

and there are 36 girls in Math Zoom Academy.

Solution 2

Let the number of girls in Math Zoom Academy to be x. Then number of boys in Math Zoom Academy will be $x + 24$. Since the ratio of boys to girls is $5:3$ we have the following equation:

$$\frac{5}{3} = \frac{x+24}{x}.$$

Cross multiplying gives us

$$5x = 3(x+24).$$

Distributing and combining like terms we have

$$2x = 72$$

so we have

$$x = 36.$$

Therefore, there are 36 girls in Math Zoom Academy.

Problem 1.3 Solve the following questions involving percentages.

(a) The table shows some of the results of a survey. What percentage of the males surveyed read the newspaper?

	Read	Don't Read	Total
Male	?	26	?
Female	58	?	96
Total	136	64	200

Answer

75%

Solution

Since 200 people were surveyed in total and we know 96 of them were female, a total of

$$78 + 26 = 104$$

males were surveyed. Similarly, a total of 136 people surveyed read the paper. Since 58 of them are female the other

$$136 - 58 = 78$$

were male. We can divide this by the total number of males

$$78 \div 104 = 0.75 = 75\%.$$

so see that 75% of the males surveyed read the newspaper.

(b) The table below gives the percent of students in each grade at school A and school B.

	K	1	2	3	4	5	6
A	21%	12%	11%	15%	13%	17%	11%
B	18%	11%	16%	11%	13%	14%	17%

School A has 100 students, and school B has 200 students. If the two schools combined, what percent of the students are in grade 6?

Answer

15%

Solution

We know there are

$$100 + 200 = 300$$

total students. We need to know how many students are in grade 6. 11% of the 100 students in school A are in grade 6, so

$$11\% \times 100 = 11$$

students in school A are grade 6. Similarly,

$$17\% \times 200 = 34$$

students are in grade 6 from school B. This gives a total of

$$11 + 34 = 45$$

students in grade 6 if two schools combine. Therefore,

$$\frac{45}{300} = 0.15 = 15\%$$

of the students would be in grade 6 if the two schools combined.

Problem 1.4 Percent Increase and Decrease

(a) A merchant offers a large group of items at 30% off. Later, the merchant takes 20% off these sale prices and claims that the final price of these items is 50% off the original price. What is the true total discount?

Answer

44%

Solution 1

The original price of the item is not given in the question. We can assign a price for it to help us with the calculation. Let's say the original price of the item is $100. With the first 30% off, the new price is

$$(1 - 30\%) \times 100 = 0.70 \times 100 = 70$$

dollars. With another 20% off, the price becomes

$$(1 - 20\%) \times 70 = 0.80 \times 70 = 56$$

dollars. This is a total discount of

$$100 - 55 = 44$$

dollars, so dividing by the original price the percentage discount is

$$44 \div 100 = 0.44 = 44\%.$$

So the merchant's claim is not correct and the true discount is 44% off the original price.

Solution 2

More generally, we can assume x is the original price. With the first 30% off, the sales price will be $0.7x$. With the next 20% off, the sales price will be $0.8 \times 0.7x = 0.56x$. Then compare the final price $0.56x$ with the original price x, we have

$$\frac{x - 0.56x}{x} = \frac{0.44x}{x} = 0.44.$$

So, the true total discount is 44%.

(b) The number of students going on a field trip changed from 24 to 36. What was the percentage increase in the number of students going on a field trip? Later, the number of students going changed back from 36 to 24. What was the percentage decrease in the number? Give both answers as a percentage rounded to the nearest whole number.

Answer

Increase: 50%, Decrease: $\approx 33\%$

Solution

An increase from 24 to 36 students is a change of

$$36 - 24 = 12$$

students. Dividing by the original amount, this is a

$$12 \div 24 = 0.50 = 50\%$$

increase. Decreasing from 36 to 24 students is also a change of

$$36 - 24 = 12$$

students. However, the 'original' amount is now the 36 students, so the change is now a

$$12 \div 36 = 0.\overline{3} = 33.\overline{3} \approx 33\%$$

decrease.

Problem 1.5 Discounts, Coupons, and Sales Tax

(a) Samantha wanted to buy a new iPhone case that was originally priced $28, and she had a coupon for 15% off. What would be the discounted price of the iPhone case, after the coupon was applied?

Answer

$23.80

Solution

To calculate the discounted price, we calculate the discount amount by multiplying the original price by the percent discount, which is $28 \times 15\%$, and then subtract this value from the original price. This method can be simplified to the following expression and value: $28 \times (1 - 15\%) = 28 \times 0.85 = 23.80$ dollars.

(b) In the same situation as part (a), if the state tax rate was 8%, what was the final amount Samantha had to pay?

Answer

$25.70

Solution

To pay the price with sales tax added, we should calculate the tax first, and then add it to the price to pay. The discounted price was already calculated to be $23.80, all we need to do now is to add the 8% tax onto the value: $23.08 \times (1 + 8\%) = 23.80 \times 1.08 = 25.704$, so Samantha would pay $25.70 after rounding.

(c) The cashier, Roger, applied the tax first and got the amount Samantha needed to pay before Samantha showed him the coupon. Roger took the coupon and applied the 15% discount and reached a final amount for Samantha to pay. Samantha made the payment and received her new lemon-patterned iPhone case. She was very happy and put it on the phone immediately. However, on her way home, she kept wondering if she might have ended up paying more tax because the cashier applied the tax before applying the coupon. Can you help her figure it out?

Answer

The prices would be the same.

Solution

If Roger adds the tax first, the final total Samantha must pay is:

$$28 \times (1 + 8\%) \times (1 - 15\%).$$

Rather than doing this calculation, let's write the expression for how much she would pay with the discount applied first:

$$28 \times (1 - 15\%) \times (1 + 8\%).$$

By the commutative property of multiplication, these two expressions have the exact same value. Therefore the resulting prices are the same for both procedures.

Problem 1.6 When gold sold for $16 an ounce, Johnny found $6 worth of gold in his claim. Gold presently sells for $328 an ounce. How many dollars is Johnny's amount of gold worth today?

Answer

$123

Solution 1

Dividing the new price per ounce by the original price per ounce we have that the new price is

$$328 \div 16 = 20.5 = 2050\%$$

of the old price. Therefore Johnny's $6 of gold is now worth

$$2050\% \times 6 = 20.5 \times 6 = 123$$

dollars.

Solution 2

Johnny found gold originally worth $6. Since gold was $16 an ounce, this means Johnny found

$$6 \div 16 = \frac{3}{8} = 0.375$$

of an ounce of gold. Since gold is now worth $328 an ounce, Johnny's gold is worth

$$0.375 \times 328 = 123$$

dollars.

Solution 3

$$\frac{6}{16} = \frac{x}{328}.$$

Solve for x, then,

$$x = \frac{6}{16} \times 328 = 123.$$

Thus, Johnny's amount of gold is now worth $123.

Problem 1.7 John is on a business trip with his friends Eric, Nick and Oscar. Oscar lost all his money, so the friends wanted to help him. Each gave Oscar the same amount of money. However, Eric gave Oscar 20% of his own money, John gave Oscar 25% of his money, and Nick gave Oscar $\frac{1}{3}$ of his money. What percent of the group's money does Oscar now have?

Answer

25%

Solution

The problem does not say how much money each of the friends gave to Oscar. Since all the amounts are described in percentage or fractions, and the question also asks for percentage, we may assume each friend gave Oscar $100. Oscar has received $100 from each of the three friends, so he has $300 now. To answer the question, we need to figure out the total amount of money they have, and it is necessary to find out the amount of money each of the friends had before giving $100 to Oscar.

Eric gave Oscar 20% of his money. This means $100 is 20% of Eric's money, so Eric had $100 \div 20\% = 500$ dollars to begin with.

John gave Oscar 25% of his money. Thus John had $100 \div 25\% = 400$ dollars before giving the Oscar.

Nick gave Oscar $\frac{1}{3}$ of his money. Therefore Nick had $100 \div \frac{1}{3} = 300$ dollars at the beginning.

Putting all these amounts together, the total amount of money they have is $500 + 400 + 300 = 1200$ dollars.

Oscar has 300 dollars now, and this is $\frac{300}{1200} = \frac{1}{4} = 25\%$ of the group's money.

Problem 1.8 Henry reads 160 pages of a book per day. After 5 days, Henry has $\dfrac{3}{5}$ of the book remaining. How many pages does the book have?

Answer

2000

Solution

Henry reads 160 pages per day, so in 5 days he reads

$$5 \times 160 = 800$$

pages. Since Henry has $\dfrac{3}{5}$ of the book remaining the 800 pages he has read are $\dfrac{2}{5}$ of the book. Therefore, we can divide to get that the book has

$$800 \div \frac{2}{5} = 2000$$

pages in total.

Problem 1.9 Sixty percent of the people on a subway train are seated. As some people prefer standing, only 75% of the seats on the subway are filled. If there are 12 empty seats, how many people are on the train?

Answer

60

Solution 1

Since 75% of the seats are filled 25% of the seats are empty. Since there are a total of 12 empty seats, there must be

$$12 \div 25\% = 12 \div 0.25 = 48$$

seats in total. 75% of these seats are filled, so

$$75\% \times 48 = 0.75 \times 48 = 36$$

seats are filled. These 36 seats are filled by 36 people, and since 60% of the total people are seated, there must be a total of

$$36 \div 60\% = 36 \div 0.6 = 60$$

people on the train.

Solution 2

We know 75% of the seats are filled, so 25% of the seats are empty. Therefore the ratio of filled to empty seats is

$$75\% : 25\% = 75 : 25 = 3 : 1.$$

Since there are 12 empty seats, we multiply both sides of this ratio by 12 to get

$$3 : 1 = 36 : 12$$

so there are 36 filled seats. Since 60% of the people are seated on the train, 40% are standing. Therefore the ratio of seated to standing people is

$$60\% : 40\% = 60 : 40 = 3 : 2.$$

We know there are 36 seated people, so multiplying the ratio by

$$36 \div 3 = 12$$

we have

$$3 : 2 = 36 : 24$$

so there are 24 standing people. Thus, there are a total of

$$36 + 24 = 60$$

people on the train.

Problem 1.10 A fruit salad consists of crabapples, cranberries, black berries, and black cherries. If there are twice as many cranberries as crabapples and three times as many black berries as black cherries and four times as many black cherries as cranberries and the fruit salad has 280 total fruits, then how many black cherries does it have?

Answer

64

Solution 1

To stay organized, let's list out all the ratios between each two fruits using the abbreviations CB = cranberries, CA = crabapples, BB = black berries, and BC = black cherries.

$$CB : CA = 2 : 1, BB : BC = 3 : 1, BC : CB = 4 : 1.$$

We can then find the ratios between all the fruits. We know the ratio of $CB : CA$ is $2 : 1$, and the ratio of $BC : CB$ is $4 : 1 = 8 : 2$, so the ratio $BC : CA$ is $8 : 1$ and the ratio of the three $CB : CA : BC = 2 : 1 : 8$. Similarly, since the ratio of $BB : BC$ is $3 : 1 = 24 : 8$ we have $CB : CA : BC : BB = 2 : 1 : 8 : 24$.

Since

$$2 + 1 + 8 + 24 = 35$$

the ratio above tells us that

$$\frac{8}{35}$$

of the fruits in the fruit salad are black cherries. Since the fruit salad has 280 total fruits, there are

$$\frac{5}{35} \times 280 = 64$$

black cherries in the fruit salad.

Solution 2

There are less crabapples than cranberries, less cranberries than black cherries, and less black cherries than black berries. Hence there are less crabapples than any other fruit. Suppose there was 1 crabapple in the fruit salad. Since there are twice as many cranberries, there would be

$$1 \times 2 = 2$$

cranberries. There are four times as many black cherries as cranberries, so there would be

$$2 \times 4 = 8$$

black cherries. Lastly, there are three times as many black berries as black cherries, so there would be

$$4 \times 3 = 24$$

black berries. Thus for every 1 crabapple there are 2 cranberries, 8 black cherries, and 24 black berries, a total of

$$1 + 2 + 8 + 24 = 35$$

fruit. Since

$$280 \div 35 = 8$$

if we multiple the amount of each fruit we have by 8 we will have the necessary 280 fruits in the fruit salad. In this case we will have

$$8 \times 8 = 64$$

black cherries.

2   Solutions to Chapter 2 Examples

Problem 2.1 Ratios warmup problems!

(a) In a far-off land three fish can be traded for two loaves of bread and one loaf of bread can be traded for six ears of corn. How many ears of corn are worth the same as one fish?

Answer

4

Solution

The ratio of fish to bread is $3:2$ and the ratio of bread to corn is $1:6$ in worth. Since

$$1:6 = 2:12$$

we can trade two loaves of bread for 12 ears of corn. Therefore, starting with twelve ears of corn, we can trade for two loaves of bread which we can then trade for three fish. Therefore, the ratio of corn to fish is $12:3$. Dividing by 3 we have

$$12:3 = 4:1$$

so four ears of corn are worth the same as one fish.

(b) The ratio of llamas to ostriches in the Math Zoom Academy petting zoo is $4:7$. If there are total of 44 llamas and ostriches in the petting zoo, how many of the them are llamas?

Answer

16

Solution 1

Since the ratio of llamas to ostriches is $4:7$, 4 out of every

$$4+7 = 11$$

are llamas. Thus, $\frac{4}{11}$ of the 44 total of llamas and ostriches are llamas. Therefore, the petting zoo has

$$\frac{4}{11} \times 44 = 16$$

llamas.

Solution 2

(Algebra) Let the number of llamas be x, so the number of ostriches is $44 - x$. Since the ratio of llamas to ostriches is $4 : 7$ we have the equation

$$\frac{4}{7} = \frac{x}{44 - x}.$$

Cross-multiplying we have

$$4 \times (44 - x) = 7x$$

so distributing and combining like terms we have

$$176 = 11x.$$

We then can solve for x to get

$$x = \frac{176}{11} = 16,$$

so there are 16 llamas at the petting zoo.

Problem 2.2 Use percentages to solve the following problems.

(a) Three bags of jelly beans contain 26, 28, and 30 beans. The bags consist of respectively 50%, 25%, and 20% yellow jelly beans. All three bags of beans are dumped into one bowl. What percent of all beans are yellow jelly beans? Round your answer to the nearest percent.

Answer

$\approx 31\%$

Solution

First find the number of yellow jellybeans in each bag. The first bag has

$$50\% \times 26 = 13,$$

the second has

$$25\% \times 28 = 7,$$

and the third has

$$20\% \times 30 = 6$$

yellow jelly beans. So the total number of yellow jelly beans in the three bags is

$$13 + 7 + 6 = 26.$$

The number of jelly beans of all colors is

$$26 + 28 + 30 = 84$$

so the percent of yellow jellybeans is

$$26 \div 84 = \frac{13}{42} \approx 0.3095 \approx 31\%.$$

(b) Jong-Zhi took a math test that had 12 arithmetic questions, 15 algebra questions and 18 geometry questions. She got 75% of the arithmetic questions correct and 60% of the algebra questions correct. How many of the geometry questions must she get correct to get a passing grade of 75%?

Answer

16

Solution

The test in total has

$$12 + 15 + 18 = 45$$

questions. For Jong-Zhi to get 75% of them correct, she needs to get

$$75\% \times 45 = 0.75 \times 45 = 33.75$$

questions correct. Thus, to pass Jong-Zhi must get at least 34 questions correct. We know she got 75% of the arithmetic questions correct. Since there are 12 questions, she got

$$75\% \times 12 = 0.75 \times 12 = 9$$

arithmetic questions correct. Similarly, she got 60% of the 15 algebra questions correct which is an additional

$$60\% \times 15 = 9$$

correct questions. Therefore Jong-Zhi has answered

$$9 + 9 = 18$$

questions correct so far. Therefore she needs to answer

$$34 - 18 = 16$$

of the 18 geometry questions to get a passing grade on the test.

Problem 2.3 Business is a little slow at Lou's Fine Shoes, so Lou decides to have a sale. On Friday, Lou increases all of Thursday's prices by 10%. Over the weekend, Lou advertises the sale: "Ten percent off the list price. Sale starts Monday." How much does one pair of shoes cost on Monday that cost $40 on Thursday?

Answer

$39.60

Solution

On Friday, the new price of the $40 shoes will be 10% higher than Thursday which is

$$(1 + 10\%) \times 40 = 1.10 \times 40 = 44$$

dollars. Then on Monday, the price is 10% of the $44 price, which is

$$(1 - 10\%) \times 44 = 0.90 \times 40 = 39.60.$$

Therefore, the shoes cost $39.60 on Monday.

Problem 2.4 In the popular TV show "Who Wants to be a Millionaire", contestants earn certain amount of money based on the number of questions they answer correctly. The dollar values of each question are shown in the following table (where k = 1000).

Question	1	2	3	4	5	6	7	8
Value	100	200	300	500	1k	2k	4k	8k
Question	9	10	11	12	13	14	15	
Value	16K	32K	64K	125K	250K	500K	1000K	

Between which two questions is the percentage increase of the value the smallest?

Answer

2 and 3

Solution

We start finding the percentage increase between questions. Between questions 1 and 2 there is a change of

$$200 - 100 = 100$$

dollars. Dividing by the original dollar amount (for question 1), we get that this is an increase of

$$100 \div 100 = 1 = 100\%.$$

Between questions 2 and 3 there is also a change of

$$300 - 200 = 100$$

dollars, which now is a percentage increase of

$$100 \div 200 = 0.50 = 50\%.$$

Using the same method between questions 3 and 4, we get a change of

$$500 - 300 = 200$$

divided by the original

$$200 \div 300 = 0.\overline{6} = 66.\overline{6}\%,$$

so there is roughly a 67% increase between questions 3 and 4. For the rest of the questions, the dollar value at least doubles (it doubles for every question except 11 to 12, where it more than doubles). Since all of these are at least a 100% increase, we get that the increase between questions 2 and 3 has the smallest percentage increase among all the questions.

Problem 2.5 Jim is paid an 8% commission on the first $800 of weekly sales, and a 14% commission on any sales past $800. If Jim's sales were $1300, what was his commission?

Answer

$134

Solution

Jim earns 8% commission on the first $800 of sales, a total of

$$8\% \times 800 = 0.08 \times 800 = 64$$

dollars. He earns 14% on the additional

$$1300 - 800 = 500$$

dollars of sales, for an additional commission of

$$14\% \times 500 = 0.14 \times 500 = 70$$

dollars. In total, Jim's earns

$$64 + 70 = 134$$

dollars as commission.

Problem 2.6 Vitamin tablets are packed in three different sized bottles: small (S), medium (M) and large (L). The medium size costs 50% more than the small size and contains 20% fewer tablets than the large size. The large size contains twice as many tablets as the small size and costs 30% more than the medium size. Rank the three sizes from best to worst buy.

Answer

Best: M, Middle: L, Worst: S

Solution

The question asks to find out which size is the best buy and which is the worst. This ranking is best obtained by calculating the unit price for each tablet.

In terms of the price, we can assume the price of the small package is 100 cents. The medium package costs 50% more than the small one, so the price of medium package is

$$(100\% + 50\%) \times 100 = 1.50 \times 100 = 150$$

cents. The large package costs 30% more than the medium one, so the price of the large package is

$$(100\% + 30\%) \times 150 = 1.30 \times 150 = 195$$

cents.

Now that we have the prices determined, let us work on the amounts of vitamins each packaging contains. Again, assume the small package contains 100 tablets. The large package contains twice as much tablets as the small one, so the large package contains

$$100 \times 2 = 200$$

tablets. The medium package contains 20% fewer tablets than the large one, so the medium package contains

$$(100\% - 20\%) \times 200 = 0.80 \times 200 = 160$$

tablets.

By now we have the prices and numbers of tablets for each packaging, we can calculate the unit prices per tablet. For clarity we display the data in the following table.

Size	Price	Num. Tablets	Unit Price
Small	100	100	$100/100 = 1$
Medium	150	160	$150/160 = 0.9375$
Large	195	200	$195/200 = 0.975$

From the table we see that the medium size package has the best unit price, thus it is the best buy. The large size one is the second best, and small size one is the worst buy.

Problem 2.7 Two 600 ml pitchers contain orange juice. One pitcher is 30% full and the other pitcher is 40% full. Water is added to fill each pitcher completely, then both pitchers are poured into one large container. What percent of the mixture in the large container is orange juice?

Answer

35%

Solution 1

First we can find the amount of orange juice in each pitcher. The first pitcher contains

$$30\% \times 600 = 180$$

ml and the second contains

$$40\% \times 600 = 240$$

ml of orange juice. Then the total amount of orange juice is

$$180 + 240 = 420$$

ml. Since both pitchers are filled with water completely, then the total amount of liquid in the two pitchers is

$$600 + 600 = 1200$$

ml. Therefore the mixture in the large container is

$$420 \div 1200 = 0.35 = 35\%$$

orange juice.

Solution 2

If we fill both pitchers with water, the first is 30% orange juice and the second is 40% orange juice. Since both pitchers have the same size, the percentage of orange juice after mixing is the average of the two pitchers. Therefore the large contain contains

$$(30\% + 40\%) \div 2 = 35\%$$

orange juice.

Problem 2.8 Phil Lanthropist won a great deal of money in a contest. He gave 20% of his winnings to his parents, gave 25% of the remaining money to his children, and gave the remaining $900,000 to his favorite charity. What was the total number of dollars that Phil won?

Answer

$1,500,000

Solution 1

It is easiest to work backwards. He ends by giving $900,000 to charity. This was Phil's remaining money after he gave 25% of the money he had left to charity. Hence the $900,000 was

$$100\% - 25\% = 75\%$$

of the remaining money. Therefore he had

$$900,000 \div 75\% = 900,000 \div 0.75 = 1,200,000.$$

dollars before he gave his children money. Since he started by giving his parents 20% of his winnings, the $1,200,000 was

$$100\% - 20\% = 80\%$$

of his winnings. Therefore we know his winnings was

$$1,200,000 \div 80\% = 1,500,000.$$

dollars.

Solution 2

Let the amount of Phil won to be x dollars. He gives 20% to his parents, so is left with

$$100\% - 20\% = 80\%$$

of his money. He then gives 25% of the remaining to his children, which is

$$100\% - 25\% = 75\%$$

of his remaining money. Therefore his remaining money is

$$75\% \times (80\% \times x)) = 0.75 \times 0.80 \times x = 0.6x.$$

He gives this remaining money to charity. Since he gave $900,000$ to charity we know

$$0.6x = 900000$$

so solving for x we have

$$x = \frac{900000}{0.6} = 1500000.$$

Hence Phil won $1,500,000$ in the contest.

Problem 2.9 At a party there are only single women and married men with their wives. 40% of the women are single. What percentage of the people in the room are married men?

Answer

37.5%

Solution

For convenience assume 100 women are at the party. Therefore

$$40\% \times 100 = 0.40 \times 100 = 40$$

of the women are single and

$$100 - 40 = 60$$

women are married. Thus there must also be 60 married men at the party. There are no single men at the party, so the total number of people is

$$40 + 60 + 60 = 160.$$

Hence the percentage of married men is

$$60 \div 160 = 3 \div 8 = 0.375 = 37.5\%$$

at the party.

Problem 2.10 In the fish tank at Albert's house, $\dfrac{1}{4}$ of the fish are red and the number of black fish is $\dfrac{3}{5}$ of the number of red fish. There are 24 additional fish that are all spotted. How many red fish are there?

Answer

10

Solution 1

Since $\dfrac{1}{4}$ of the fish are red and the number of black fish is $\frac{3}{5}$ that of the number of red fish,

$$\frac{1}{4} \times \frac{3}{5} = \frac{3}{20}$$

of the total fish are black. Thus,

$$\frac{1}{4} + \frac{3}{20} = \frac{8}{20} = \frac{2}{5}$$

of the fish are either black or red. Hence, $\dfrac{3}{5}$ of the fish are spotted. There are 24 spotted fish, hence the total number of fish in the tank is

$$24 \div \frac{3}{5} = 40.$$

Recalling that $\dfrac{1}{4}$ of the fish are red we know there are

$$40 \times \frac{1}{4} = 10$$

red fish.

Solution 2

Let the total number of fish in the tank to be x. One-fourth are red, so there are

$$\frac{1}{4} \times x = \frac{x}{4}$$

red fish. There are $\dfrac{3}{5}$ths as many black fish, so there are

$$\frac{3}{5} \times \frac{x}{4} = \frac{3x}{20}$$

black fish. The only other type of fish is spotted. Thus, since there are 24 spotted fish we have

$$\frac{x}{4} + \frac{3x}{20} + 24 = x$$

Combining like terms we have

$$24 = \frac{3x}{5}$$

so we can solve for x to get

$$x = 24 \times \frac{5}{3} = 40,$$

the total number of fish. One-fourth are red, so the tank contains

$$\frac{1}{4} \times 40 = 10$$

red fish.

3   Solutions to Chapter 3 Examples

Problem 3.1 Solve the following.

(a) A school bought some basketballs and volleyballs, 50 balls in total. There are 10 more basketballs than volleyballs. How many balls of each kind are there?

Answer

Volleyballs: 20; Basketballs: 30

Solution

This time we are working with bigger numbers, and we don't want to draw 50 balls to figure out how many of each there are. Instead we will draw some long bars that will represent the number of balls of each kind that we have. The bar for the number of basketballs will be longer because we have 10 more basketballs than volleyballs:

Basketballs	
Volleyballs	10

$= 50$ balls

If we had 10 more Volleyballs, we would have the same amount of balls of each and we would have $50 + 10 = 60$ balls in total.

$= 60$ balls

So, if we count the Basketballs alone, we should have $60 \div 2 = 30$ of them! Which leaves us with $30 - 10 = 20$ Volleyballs.

(b) When Matt was 17 years old, Nathan was 23. This year the sum of their ages is 50. What are their ages this year?

Answer

Matt; 22; Nathan: 28

Solution

The difference of their ages is the same every year, so Nathan is $23 - 17 = 6$ years older than Matt. This year their ages look like

Nathan	6
Matt	

$= 50$ years

If Matt was 6 years older, they would have the same age and the sum of their ages would be 56 years

Nathan	6

$= 56$ years

So, Nathan is $56 \div 2 = 27$ years old, and Matt is $27 - 6 = 21$ years old.

(c) David and Ed went to pick cherries. Together they picked 82 cherries. If David gave 4 cherries to Ed, they would have the same number of cherries. How many cherries did each of them pick?

Answer

Ed: 37; David: 45

Solution

Let's see how the number of cherries looks like if David gives 4 cherries to Ed:

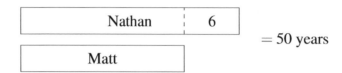

We can see that if David had 8 less cherries, they would have $82 - 8 = 74$ cherries in total and then they would both have the same amount of cherries.

So, Ed has $74 \div 2 = 37$ cherries and David has $37 + 8 = 45$ cherries.

Problem 3.2 Solve the following problems.

(a) Alice, Beth, and Cynthia have $50 in total. Alice has twice as much money as Beth, and Beth has 3 times as much as Cynthia. How much money does each have?

Answer

Alice: $30; Beth: $15; Cynthia: $5

Solution

Alice has more money than Beth, and Beth has more money than Cynthia, so Cynthia is the one that has the least amount of money and Alice is the one that has the most money. We can think that Alice, Beth and Cynthia have all their money in $1 bills, so they have exactly 50 bills. For each bill that Cynthia has, Beth will have two, and for each bill that Beth has, Alice will have three. So, for each bill that Cynthia has, Alice will have $2 \times 3 = 6$ bills. Take some empty envelopes an put in each of them 1 bill for Cynthia, 3 bills for Beth, and 6 bills for Alice. Each envelope then has $1 + 3 + 6 = 10$ bills. Since we have 50 bills in total, we will have $50 \div 10 = 5$ envelopes with $10 each. This way, Cynthia has 5 dollars, Beth has $3 \times 5 = 15$ dollars and Alice has $5 \times 6 = 30$ dollars.

(b) Allison is 14 years old, and her dad is 50 years old. How many years ago was Allison's dad's age 5 times Allison's age?

Answer

5 years ago

Solution

The difference of Allison's age and her dad's age is $50 - 14 = 36$ years. Note that it is the same difference every year. Suppose we are now on the year when her dad's age is 5 times her age. We know that the difference of their ages is 36, and the ratio of their ages is 5, so Allison's age that year was $36 \div (5 - 1) = 9$. Since she is now 14 years old, that happened $14 - 9 = 5$ years ago.

Problem 3.3 Answer the following.

(a) In the orchard there are 144 trees, which are either apple trees or peach trees. If there were 12 fewer apple trees and 20 more peach trees, then the numbers of the two kinds of trees would have been the same. How many trees of each kind are there?

Answer

Peach trees: 56; Apple trees: 88

Solution

Let's take a look at how many trees of each kind we have:

If we indeed had 12 fewer apple trees and 20 more peach trees, we would have instead $144 - 12 + 20 = 152$ trees in total

and so we would have $152 \div 2 = 76$ of each. This means that we actually have $76 + 12 = 88$ apple trees and $76 - 20 = 56$ peach trees.

(b) Mary, Rose and Catherine are working on a school project and have some big pieces of ribbon of different sizes. If they glue together their pieces of ribbon they get a ribbon that is 840 centimeters long. Their teacher measured each of their ribbons and told them that Rose's ribbon is 130 centimeters longer than Mary's, and Catherine's is 220 centimeters longer than Rose's. Can you figure out how long is each ribbon?

Answer

Mary: 120 cm; Rose: 250 cm; Catherine: 470 cm

Solution

Look at the pieces of Ribbon of each of the girls. Catherine has the longest piece of ribbon!

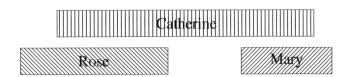

Let's cut the ribbons in smaller pieces! Take Catherine's ribbon and cut out of it 220 centimeters. So Catherine's ribbon is made of a piece as long as Rose's and one more that is 220 centimeters long.

Now take those Rose pieces and cut 130 centimeters out of each. From each you get a piece that is as long as Mary's and an extra piece that is 130 long.

We know that all the pieces together would make up 840 centimeters, so those Mary pieces together will be
$$840 - 130 - 130 - 220 = 360$$
centimeters long, which means each Mary piece must be $360 \div 3 = 120$ centimeters long! And we also know that Rose's ribbon is $120 + 130 = 250$ centimeters long and Catherine's is $250 + 220 = 470$ centimeters long!

Problem 3.4 In a school, there are 165 students in the third and fourth grades altogether. If there were 6 more third graders, the third grade would have twice as many students as the fourth grade. How many students are there in each of the two grades?

Answer

4th grade: 57; 3rd grade: 108

Solution

If we had 6 more third graders, we would have $165 + 6 = 171$ students in total. In that case, for each fourth grader we would have 2 third graders and we would be able to make $171 \div (1+2) = 57$ groups of 3 students. That would mean that we had 57 fourth graders and $57 \times 2 = 114$ third graders. Now, since we were assuming we had 6 extra students in the third grade, we actually have $114 - 6 = 108$ third graders and 57 fourth graders.

Problem 3.5 Coach just brought a *huge* bag with 75 balls for gym class. He told us that in the bag there are twice as many basketballs as soccer balls, and that there are 3 more volleyballs than soccer balls. He won't let us play with them until we figure out exactly how many of each he brought. Can you help us out?

Answer

Soccer balls: 18; Volleyballs: 21; Basketballs: 36

Solution

If we got rid of the 3 *extra* volleyballs, we would have $75 - 3 = 72$ balls in total and we would have the *same* number of soccer balls and volleyballs. Since we know that we have 2 basketballs for each soccer ball, we can make groups of 4 balls: 1 soccer ball, 1 volleyball and 2 basketballs. So, we have exactly $72 \div 4 = 18$ groups of 4 balls. That means we have 18 soccer balls, 18 volleyballs and $18 \times 2 = 36$ basketballs. Now we just need to put back the 3 volleyballs we pretended we didn't have, and so, we actually have $18 + 3 = 21$ volleyballs.

Problem 3.6 I just found a box with Christmas ornaments in the attic. It has a huge label on the front that says it contains 200 Christmas ornaments of three different colors (red, blue, and white). I'm so clumsy that I broke some of them by accident. It seems I broke 14 white ornaments and now I have the same number of white ornaments and blue ornaments. Mom says that before I broke them we had 4 more red ornaments than 3 times the number of white ornaments. How many red ornaments were there in the box before I found it?

Answer

White: 42; Blue: 28; Red: 130

Solution

Let's rewind time and pretend no baubles are broken. Also, let's put aside 4 red baubles so the number of red baubles is exactly the same as 3 times the number of white baubles. Since we have the number of red baubles is now a multiple of the number of white baubles, it would help us a lot if we could say something similar about the blue baubles. Let's pretend we had 14 more blue baubles, so we would have $200 + 14 - 4 = 210$ baubles in total, and we would have the *same* number of white and blue baubles. So, if there were no broken baubles, we would have $210 \div (1 + 1 + 3) = 42$ white baubles, $42 - 14 = 28$ blue baubles, and $42 \times 3 + 4$ red baubles.

Problem 3.7 David has 3 times as much money as Chris. If David spends \$240, and Chris spends \$40 dollars, they will have the same amount of money. How much money do each of them have originally?

Answer

Chris: \$100; David: \$300

Solution

The difference between the amount of money David and Chris have is $240 - 40 = 200$ dollars. We can do something similar as in the previous problems to figure out how much money each of them have. We will take empty envelopes and put in them some \$1 bills. This time we know that for each bills Chris has, David will have 3. The one thing that will be different in this problem as in the previous ones, is that this time we know the *difference* of the amounts of money they have, so in each envelope we need to put 3 bills for David and *remove* 1 bill for Chris, so we will have $3 - 1 = 2$ bills in each envelope. We can make $200 \div 2 = 100$ envelopes like this, so Chris has 100 dollars and David has 100×3 dollars.

Problem 3.8 The average price of a basketball, a soccer ball, and a volleyball is \$36. The basketball is \$10 more expensive than the volleyball, the soccer ball is \$8 more expensive than the volleyball. How much is the soccer ball?

Answer

\$38

Solution

Since the average price of the balls is 36 dollars, the *sum* of their prices is $36 \times 3 = 108$ dollars.

Volleyball		
Basketball	10	= \$108
Soccer ball	8	

Note that if we take out those 8 and 10 portions on the Basketball and Soccer ball bars, we would have three bars the same size as the Volleyball bar:

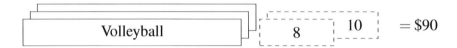

So, the price of *one* volleyball is $90 \div 3 = 30$ dollars. That means that the price of a Soccer ball is $30 + 8 = 38$ dollars.

Problem 3.9 Captain Hook, Popeye, and Sinbad went out to the sea to hunt for treasure. They all found diamonds. The total number of diamonds found by Captain Hook and Popeye is 80. The total number of diamonds found by Popeye and Sinbad is 70. The total number of diamonds found by Captain Hook and Sinbad is 50. How many diamonds did each of them find?

Answer

Captain Hook: 30; Popeye: 50; Sinbad: 20

Solution

The problem is telling us how many diamonds they got by pairs. Let's use bars to represent the number of diamonds each of them found.

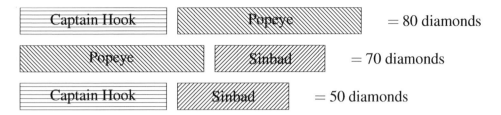

If we group all of these bars together, we have 2 bars of each and a total of 200 diamonds altogether.

So, if we had only one of each bar, we would have $200 \div 2 = 100$ diamonds. Note that if we get rid of two of the bars, we would know how many diamonds the third person found:

		Sinbad	= 20 diamonds
Captain Hook			= 30 diamonds
	Popeye		= 50 diamonds

So, Sinbad found $100 - 80 = 20$ diamonds, Captain Hook found $100 - 70 = 30$ diamonds and Popeye found $100 - 50 = 100$ diamonds.

Problem 3.10 In the morning, there were 149 vehicles (cars and trucks) in the parking lot. In the afternoon 8 more cars entered the parking lot, and 5 trucks drove away. It turns out that 3 times the number of cars in the parking lot is the same as 5 times the number of trucks. How many cars and how many trucks were there in the parking lot in the morning?

Answer

Cars: 87; Trucks: 62

Solution

8 cars came in the parking lot and 5 trucks left the parking lot, so we have now $149 + 8 - 5 = 152$ vehicles in the parking lot. As we know *"3 times the number of cars is the same as 5 times the number of trucks"*, we have that for every 5 cars we have 3 trucks. We can make $152 \div (3 + 5) = 19$ groups of 8 vehicles, so we have $19 \times 5 = 95$ cars and $19 \times 3 = 57$ trucks. This means that in the morning we had $95 - 8 = 87$ cars and $57 + 5 = 62$ trucks.

4   Solutions to Chapter 4 Examples

Problem 4.1 There are some chickens and some rabbits on a farm. Suppose there are 50 heads and 140 feet in total among the chickens and rabbits, how many chickens are there? How many rabbits?

Answer

20 rabbits, 30 chickens.

Solution 1

First of all, each animal has one head. A chicken has two feet, and a rabbit has four feet. There are 50 heads, so there are 50 animals.

Now assume all 50 animals are chickens, there should be total 100 feet. But there are $140 - 100 = 40$ additional feet. These 40 feet came from the rabbits which were wrongfully assumed to be chickens. Each rabbit has 2 more feet than the wrongfully assumed chicken, so we need to count in two additional feet for each rabbit. Now $40 \div 2 = 20$, which is the number of rabbits. The rest is the number of chickens, which is 30.

Solution 2

Imagine all the animals are trained with special skills for a circus performance. When the trainer blows a whistle, each chicken stands with one foot, and each rabbit stands up with two hind legs. All the animals lift half the number of feet up to the air. On the ground now, there should be $140 \div 2 = 70$ feet.

Now each chicken has exactly one foot on the ground. However many chickens we have, we should count exactly same number of feet. But each rabbit would have two feet on the ground. Therefore, the extra number of feet on the ground 70, in comparison to the number of heads given, which is 50, would be the number of rabbits: $70 - 50 = 20$.

Solution 3

Let x be the number of chickens, and y be the number of rabbits. There are 50 animals in total, so

$$x + y = 50.$$

Each chicken has 2 feet, while each rabbit has 4. Since there are 140 feet in total, we

have
$$2 \times x + 4 \times y = 140.$$

This gives us the system of equations

$$\begin{cases} x+y &= 50, \\ 2x+4y &= 140. \end{cases}$$

To solve this system of two equations, we use substitution method. The first equation can be be written as

$$x = 50 - y.$$

Substituting this into the second equation we have

$$2 \times (50 - y) + 4y = 140,$$

so distributing and combining like terms gives us

$$2y = 40.$$

Hence we can solve for y to get

$$y = \frac{40}{2} = 20.$$

Plugging this back in we also get

$$x = 50 - y = 50 - 20 = 30.$$

Thus there are 20 chickens and 30 rabbits.

Problem 4.2 Answer the following using the Chicken and Rabbit Method

(a) Mary is baking a lot of cookies for a bake sale at Samantha's school. To bake the cookies she needs a total of 60 eggs. The eggs come in small cartons containing 6 eggs or large cartons containing 12 eggs. If Mary buys a total of 7 cartons, how many small cartons and how many large cartons does she buy?

Answer

4 small, 3 large

Solution

If Mary bought 7 small cartons of eggs, she would have a total of

$$7 \times 6 = 42$$

eggs, which is

$$60 - 42 = 18$$

less than she needs. Hence some of the cartons must be large cartons. Since a large carton contains

$$12 - 6 = 6$$

extra eggs, if Mary replaces

$$18 \div 6 = 3$$

of the small cartons with large cartons she will have the correct number of eggs. Therefore Mary buys

$$7 - 3 = 4$$

small cartons of eggs and 3 large cartons of eggs.

(b) Morgan needed 70 sticks for a project at school. Each stick is either 3 inches or 5 inches and the total length of all the sticks combined is 270 inches. How many 3 inch sticks and 5 inch sticks are there?

Answer

40 three inch, 30 five inch sticks.

Solution 1

If all the sticks were 3 inch sticks, the total length of the 70 sticks would be

$$70 \times 3 = 210$$

inches. Since the actual length is

$$270 - 210 = 60$$

inches longer, some of the 3 inch sticks need to be replaced with 5 inch sticks. Each 5 inch stick is

$$5 - 3 = 2$$

inches longer. Since the actual length is 60 more than the length of all 3 inch sticks, we must replace

$$60 \div 2 = 30$$

3 inch sticks with 5 inch sticks. Hence there are

$$70 - 30 = 40$$

three inch sticks and 30 five inch sticks.

Solution 2

(Algebra) Let x be the number of 3 inch sticks and y be the number of 5 inch sticks. There are 70 sticks in total, so

$$x + y = 70.$$

The total combined length of the 3 and 5 inch sticks is 270 so we also have

$$3 \times x + 5 \times y = 260.$$

This gives the system of equations

$$\begin{cases} x + y &= 70, \\ 3x + 5y &= 270. \end{cases}$$

Multiplying the first equation by 3 we get

$$3x + 3y = 210$$

and subtracting this from the second equation we have

$$2y = 60.$$

Hence solving for x we have

$$y = 30,$$

the number of 5 inch sticks. Lastly, substituting back into the first equation,

$$x + 30 = 70$$

so

$$x = 70 - 30 = 40,$$

the number of 3 inch sticks.

Problem 4.3 The counselor brought his 51 students to the lake to go rowing, 6 people for each big boat and 4 people for each small boat. They rented 11 boats to fit everyone with no empty seats. How many big and small boats each did they rent?

Answer

4 big boats, 7 small boats

Solution 1

Remember that the counselor is a person too! This means there are a total of 52 people who fit in the boats. With 11 small boats, there is only room for

$$11 \times 4 = 44$$

people, so the counselor must have rented some big boats. The counselor needs room for

$$52 - 44 = 8$$

more people. Since a big boat carries

$$6 - 4 = 2$$

more people than a small boat, switching

$$8 \div 2 = 4$$

of the small boats to big boats will make sure the 11 boats fit everyone. This means that the counselor rented 4 big boats and

$$11 - 4 = 7$$

small boats.

Solution 2

(Algebra) Let x be the number of big boats and y be the number of small boats that the counselor rented. We know there are 11 boats in total, so

$$x + y = 11.$$

We need room for 51 students and 1 counselor, so 52 people in total. Each big boat carries 6 and each small boat carries 4, so

$$6 \times x + 4 \times y = 52.$$

We then need to solve the system of equations

$$\begin{cases} x+y & = & 11, \\ 6x+4y & = & 52. \end{cases}$$

Multiplying the first equation by 4 we have

$$4x+4y = 44.$$

Subtracting this equation from the second equation gives us

$$2x = 52 - 44 = 8$$

so solving for x we have

$$x = 4.$$

We can then substitute into the first equation to get

$$4+y = 11$$

so solving for y we have

$$y = 11 - 4 = 7.$$

Hence we know that 4 big boats and 7 small boats were rented.

Problem 4.4 Sasha takes a mathematics competition. There are a total of 20 problems. For each correct answer, competitors receive 5 points. For each wrong answer, they instead get 1 point taken away. Sasha has 64 points in total. How many problems did she answer correctly?

Answer

14

Solution 1

If Sasha answered all 20 questions correct she would receive a perfect score of

$$20 \times 5 = 100.$$

She actually has only 64 points, so

$$100 - 64 = 36$$

points were deducted. Since 1 point is taken away for each wrong answer, each wrong answer actually decreases the score by

$$1 + 5 = 6$$

points. The number of wrong answer is then

$$36 \div 6 = 6,$$

and hence

$$20 - 6 = 14$$

is the number of questions Sasha answered correctly.

Solution 2

(Algebra) Let x be the number of correct answers and y be the number of wrong answers. There are 20 questions in total, each correct or wrong so

$$x + y = 20.$$

The final score of 64 is a combination of 5 points for each correct answer and losing one point for each incorrect answer, so

$$5 \times x - 1 \times y = 64.$$

This gives the system of equations:

$$\begin{cases} x + y & = & 20, \\ 5x - y & = & 64. \end{cases}$$

Adding these two equations we have that

$$6x = 84$$

so dividing by 6 gives us

$$x = \frac{84}{6} = 14,$$

the number of questions Sasha got correct.

Problem 4.5 Solve the following.

(a) Teachers and students from the Areteem Summer Camp visited the museum. They bought a total of 99 tickets for 218 dollars. If each teacher ticket costs 4 dollars, and each student ticket costs 2 dollars, how many teachers and students were there respectively?

Answer

10 teachers, 89 students

Solution 1

If instead we have 99 unsupervised students with no teachers, the 99 tickets would cost a total of

$$99 \times 2 = 198$$

dollars. This is

$$218 - 198 = 20$$

dollars cheaper than the actual 99 tickets. If we remove one student ticket and add one teacher ticket, it costs $2 extra. Since $20 extra was paid over the price of 99 student tickets, there must be

$$20 \div 2 = 10$$

teachers. Hence there are

$$99 - 10 = 89$$

students.

Solution 2

(Algebra) Let x be the number of teachers and y be the number of students who visit the Amusement Park. 99 tickets are bought, so

$$x + y = 99.$$

A total of $218 is spent buying these tickets. Each teacher ticket costs $4 and each student tickets costs $2, so

$$4 \times x + 2 \times y = 218.$$

We then need to solve the system of equations:

$$\begin{cases} x + y & = & 99, \\ 4x + 2y & = & 218. \end{cases}$$

Doubling the first equation we have

$$2x + 2y = 198.$$

Subtracting this from the second equation we have

$$2x = 20,$$

so we solve for x to get

$$x = \frac{20}{2} = 10.$$

Substituting into the first equation,

$$10 + y = 99$$

so

$$y = 99 - 10 = 89.$$

There are thus 10 teachers and 89 students who go to the Amusement Park.

(b) A group of 68 people rent 24 motorcycles of two kinds at a racetrack. The first kind has a capacity of 2 people and costs $40 per motorcycle. The second has a capacity of 3 people and costs $30 per motorcycle. The 68 people exactly fill all vehicles. What is the total cost in renting the 24 motorcycles?

Answer

$760

Solution 1

If all 24 motorcycles are the first kind with a capacity of 2 people, the motorcycles will fit

$$24 \times 2 = 48$$

people, meaning that

$$68 - 48 = 20$$

people will left out. The second kind of motorcycle can hold

$$3 - 2 = 1$$

extra person, so if we switch 20 of the motorcycles to the second kind everyone will fit. Hence we have

$$24 - 20 = 4$$

motorcycles of the first kind and 20 of the second. Since the first kind costs $40 per motorcycle and the second kind costs $30, the total cost is

$$4 \times 40 + 20 \times 30 = 760$$

dollars to rent the 24 motorcycles.

Solution 2

(Algebra) Let x be the number of motorcycles of the first kind and y the number of the second kind that are rented. 24 motorcycles are rented in total, so

$$x + y = 24.$$

The 68 people all fit in these 24 motorcycles, with each motorcycle holding 2 or 3 people, so

$$2 \times x + 3 \times y = 68.$$

This gives the system of equations

$$\begin{cases} x + y & = & 24, \\ 2x + 3y & = & 68. \end{cases}$$

Doubling the first equation we have

$$2x + 2y = 48,$$

so subtracting this from the second equation we have

$$y = 20,$$

the number of motorcycles of the second kind. To find the number of the first kind, we substitute $y = 20$ back into the first equation to get

$$x + 20 = 24$$

so

$$x = 4,$$

the number of motorcycles of the first kind. These motorcycles each cost either \$40 or \$30 to rent. Therefore, the total cost is

$$4 \times 40 + 20 \times 30 = 760$$

to rent the motorcycles.

Problem 4.6 Jack went hiking. His uphill speed was 3 miles per hour and downhill speed was 5 miles per hour, and he hiked a total of 6 hours including both uphill and downhill, with total distance 23 miles. How many hours did he spent uphill and downhill respectively?

Answer

3.5 hours uphill, 2.5 hours downhill

Solution 1

If we suppose Jack walk for the whole 6 hours at a speed of 3 miles per hour, he would travel a total of

$$3 \times 6 = 18$$

miles, which is

$$23 - 18 = 5$$

miles less than he actually hiked. Every hour Jack spends hiking downhill at 5 miles per hour allows him to travel an additional

$$5 - 3 = 2$$

miles than he would hiking uphill. Therefore, Jack must spend

$$5 \div 2 = 2.5$$

hours hiking downhill. Hence he spends the other

$$6 - 2.5 = 3.5$$

hours hiking uphill.

Solution 2

(Algebra) Let x be the number of hours Jack hikes uphill and y the number downhill. He hikes a total of 6 hours, so

$$x + y = 6.$$

We also know Jack's hike was 23 miles long, so as he travels 3 miles per hour uphill and 5 miles per hour downhill

$$3 \times x + 5 \times y = 23.$$

We therefore want to solve the system of equations

$$\begin{cases} x + y &= 6, \\ 3x + 5y &= 23. \end{cases}$$

Multiplying the first equation by 3 we have

$$3x + 3y = 18.$$

Subtracting this from the second equation,

$$2y = 5$$

so

$$y = \frac{5}{2} = 2.5.$$

Substituting back into the first equation

$$x + 2.5 = 6$$

we get

$$x = 6 - 2.5 = 3.5.$$

Hence Jack spends 3.5 hours hiking uphill and 2.5 hours hiking downhill.

Problem 4.7 Answer the following questions.

(a) There are some chickens and some rabbits on a farm. Suppose there are 50 heads and there are 20 more rabbit feet than chicken feet on the farm. How many chickens and how many rabbits are there?

Answer

30 chickens, 20 rabbits

Solution 1

If all 50 animals are rabbits, there are

$$50 \times 4 = 200$$

rabbit feet and 0 chicken feet, so there are 200 more rabbit feet than chicken feet. If we replace a rabbit with a chicken, we remove 4 rabbit feet and add 2 chicken feet, so the difference between rabbit feet and chicken feet decreases by

$$4 + 2 = 6.$$

Since we know there are 20 more rabbit feet than chicken feet, which is

$$200 - 20 = 180,$$

smaller than if all the animals were rabbits, we must replace

$$180 \div 6 = 30$$

rabbits with chickens. Hence there are

$$50 - 30 = 20$$

rabbits and 30 chickens on the farm.

Solution 2

Let x be the number of chickens and y be the number of rabbits. There are 50 heads, so

$$x+y=50.$$

We know there are 20 more rabbit feet than chicken feet, so

$$4\times x = 2\times y + 20.$$

Rearranging we have the system of equations

$$\begin{cases} x+y & = & 50, \\ 4x-2y & = & 20. \end{cases}$$

Doubling the first equation we have

$$2x+2y=100$$

and adding this to the second equation we have

$$6x=120.$$

Hence

$$x=\frac{120}{6}=20,$$

so substituting we have

$$20+y=50,$$

so

$$y=50-20=30.$$

Hence there are 20 rabbits and 30 chickens.

(b) The Math Club collected donations from 40 people who live in either City A or B. Each person from City A contributed \$5, and each person from City B contributed \$8. In total, \$5 more was collected from City A than from City B. How many people are there in each city?

Answer

25 in City A, 15 in City B

Solution 1

If all the people were from City A, the Math Club would have collected \$0 from City B and

$$40 \times 5 = 200$$

from City A. This is a difference of \$200 dollars. Every person who is from City B instead of City A changes this difference by

$$5 + 8 = 13$$

because there is \$5 less from City A and \$8 more from City B. If we know that \$5 more was collected from City A, which is

$$200 - 5 = 195$$

less the amount if all the people were from City A, we need to swap

$$195 \div 13 = 15$$

people from City A to City B. Hence there are

$$40 - 15 = 25$$

people from City A and 15 from City B.

Solution 2

(Algebra) Let x be the number of people from City A and y the number from City B. There are 40 donations in total, so

$$x + y = 40.$$

The club raised \$5 more on the \$5 donations from City A than on the \$8 donations from City B, so

$$5 \times x = 8 \times y + 5.$$

Rearranging we have the system of equations

$$\begin{cases} x + y &= 40, \\ 5x - 8y &= 5. \end{cases}$$

Multiplying the first equation by 5 we get

$$5x + 5y = 200$$

and subtracting the second equation from this,

$$13y = 195.$$

Solving for y,

$$y = \frac{195}{13} = 15,$$

so substituting into the first equation,

$$x + 15 = 40$$

and

$$x = 40 - 15 = 25.$$

Hence there are 25 people from City A and 15 people from City B.

Problem 4.8 More Practice with Differences.

(a) Four basketballs and five volleyballs cost 185 dollars in total. If a basketball costs 8 dollars more than a volleyball, what is the cost of one basketball?

Answer

25 dollars

Solution 1

There are a total of

$$4 + 5 = 9$$

balls bought in total. If we tried to save money and bought nine volleyballs instead of four basketballs and five volleyballs, then the price would be

$$4 \times 8 = 32$$

dollars cheaper because each volleyball is \$8 less than a basketball. Therefore 9 volleyballs cost

$$185 - 32 = 153$$

dollars in total. Hence we can fine the price of one volleyball which is

$$153 \div 9 = 17$$

dollars. Since one basketball is \$8 more, a single basketball costs

$$17 + 8 = 25$$

dollars.

Solution 2

(Algebra) Let x be the price of one volleyball. Since a basketball is \$8 more expensive, it costs

$$x+8$$

dollars. We know that four basketballs and five volleyballs cost \$185, so

$$4 \times (x+8) + 5 \times x = 185.$$

Distributing and combining like terms we have

$$9x = 163$$

so

$$x = \frac{163}{9} = 17,$$

the price of one volleyball. A basketball therefore costs

$$17 + 8 = 25$$

dollars.

(b) It requires either 45 small trucks or 36 big trucks to transport a batch of steel blocks. Given that each big truck can load 4 more tons of steel blocks than each small truck. How many tons of steel blocks are in a batch?

Answer

720

Solution 1

A full batch can be carried by 36 big trucks. Since each big truck can load 4 more tons than each small truck, if 36 small trucks are used, there is still

$$36 \times 4 = 144$$

tons of steel blocks left over. However, we know that 45 small trucks in total can carry the full batch, so the

$$45 - 36 = 9$$

extra trucks must carry the extra 144 tons. Hence each small truck can carry

$$144 \div 9 = 16$$

tons of steel blocks. Finally, this means that one batch, which is carried by 45 small trucks, is a total of

$$45 \times 16 = 720$$

tons of steel blocks.

Solution 2

The algebra method. Let the capacity of a small truck be x tons, then the capacity of each big truck is $x + 4$ tons. Since one batch is carried by 45 small trucks or by 36 big trucks, we have

$$45 \times x = 36 \times (x + 4).$$

Distributing and combining like terms we have

$$9x = 144.$$

Hence we can solve for

$$x = \frac{144}{9} = 16.$$

Be careful, as x is just the capacity of one small truck. Since one batch is carried by 45 small trucks, one batch is

$$45 \times 16 = 720$$

tons of steel blocks.

Problem 4.9 100 monks eat 100 steamed buns. If each senior monk eats 4 steamed buns, and 4 junior monks eat 1 steamed bun, how many senior monks and junior monks are there?

Answer

20 senior and 80 junior monks

Solution 1

Grouping one senior monk with four junior monks, we have a group of 5 monks that eats

$$4 + 1 = 5$$

steamed buns in total. Since there are 100 cakes in total, there must be

$$100 \div 5 = 20$$

such groups. In each group, there is only one senior monk, so there are

$$20 \times 1 = 20$$

senior monks in total. Similarly, in each group, there are 4 junior monks, so there are

$$20 \times 4 = 80$$

junior monks in total.

Solution 2

Since 4 junior monks eat 1 steamed bun, one junior monk eats

$$1 \div 4 = 0.25$$

of a steamed bun. As a senior monk eats 4 steamed buns, a senior monk eats

$$4 - 0.25 = 3.75$$

more buns than a junior monk. If all 100 monks were senior, they would eat a total of

$$100 \times 4 = 400$$

buns. However, there are only 100 steamed buns, so this is

$$400 - 100 = 300$$

buns too many. We know a senior monk eats 3.75 more buns than a junior monk, so if we change

$$300 \div 3.75 = 80$$

senior monks to junior monks, the monks will eat the correct number of steamed buns. Therefore there are 80 junior monks and

$$100 - 80 = 20$$

senior monks.

Solution 3

(Algebra) Let x be the number of senior and y be the number of junior monks. There are 100 monks in total, so

$$x + y = 100.$$

4 junior monks eat 1 steamed bun, so one junior monk eats

$$1 \div 4 = 0.25$$

buns. Since we also know a senior monk eats 4 buns,

$$4 \times x + 0.25 \times y = 100.$$

Therefore we have the system of equations

$$\begin{cases} x + y & = & 100, \\ 4x + 0.25y & = & 100. \end{cases}$$

Multiplying the second equation by 4 we get

$$16x + y = 400.$$

Subtracting the first equation from this gives us

$$15x = 300,$$

so we solve for x,

$$x = \frac{300}{15} = 20.$$

Plugging back into the first equation,

$$20 + y = 100,$$

so

$$y = 100 - 20 = 80.$$

Thus there are 20 senior monks and 80 junior monks.

Problem 4.10 Suppose there are chicken, rabbits, and sheep on a farm. There are 70 heads in total and 220 feet. If there are the same number of rabbits and sheep, how many chickens, rabbits, and sheep are on the farm?

Answer

30 chickens, 20 rabbits, 20 sheep

Solution

There are an equal number of rabbits and sheep, so pair up one rabbit with one sheep. This pair has 2 heads and 8 legs. If there are no chickens, there are

$$70 \div 2 = 35$$

such pairs, for a total of

$$35 \times 8 = 280$$

feet. This is a total of

$$280 - 220 = 60$$

extra feet. Replacing a rabbit/sheep pair with 2 chickens reduces the number of feet by

$$8 - 4 = 4.$$

Hence there are

$$60 \div 4 = 15$$

pairs of chickens and thus

$$35 - 15 = 20$$

rabbit/sheep pairs. Thus there are 30 chickens, 20 rabbits, and 20 sheep on the farm.

5   Solutions to Chapter 5 Examples

Problem 5.1 Review of the Chicken and Rabbit Method

(a) In Bob's Cycle Shop, workers received the delivery of an order Bob placed for the single seat bicycles and tricycles. There are a total of 90 seats and 215 wheels, plus all the other necessary parts and accessories. How many bicycles and how many tricycles can they assemble?

Answer

55 bicycles, 35 tricycles

Solution 1

If the workers at the shop build 90 bicycles from the parts, they will end up using all 90 seats but only

$$90 \times 2 = 180$$

wheels, so they would have

$$215 - 180 = 35$$

wheels left over. Each tricycle has 1 more wheel than a bicycle, so the workers need to turn 35 of the bicycles into tricycles. Thus they assemble

$$90 - 35 = 55$$

bicycles and 35 tricycles.

Solution 2

(Algebra) Let x be the number of bicycles and y be the number of tricycles. Each has one seat, so

$$x + y = 90.$$

Each bicycles has 2 and each tricycles has 3 wheels, so

$$2 \times x + 3 \times y = 215.$$

This gives us the system of equations

$$\begin{cases} x + y & = & 90, \\ 2x + 3y & = & 215. \end{cases}$$

Doubling the first equation we have

$$2x + 2y = 180,$$

and subtracting this from the second we have

$$y = 215 - 180 = 35.$$

Substituting this into the first equation gives us

$$x + 35 = 90$$

so we can solve for x to get

$$x = 90 - 35 = 55.$$

Thus there are 55 bicycles and 35 tricycles.

(b) There are 100 birds and cats. The birds have 80 more legs than the cats. How many birds and how many cats are there?

Answer

80 birds, 20 cats

Solution 1

There are 80 more bird legs than cat legs. Since each bird has 2 legs, if we remove

$$80 \div 2 = 40$$

birds then we are left with

$$100 - 40 = 60$$

animals and there are an equal number of bird and cat legs. Since each cat has twice as many legs as a bird, there must be half as many birds as cats. Therefore there are

$$60 \div 3 = 20$$

cats and

$$20 \times 2 = 40$$

of the remaining animals are birds. Hence there are

$$40 + 40 = 80$$

birds in total.

Solution 2

(Algebra) Let x be the number of birds and y be the number of cats. There are 100 animals in total, so

$$x + y = 100.$$

Each bird has 2 legs and each cat 4 legs, so since there are 80 more bird legs than cat legs,

$$2 \times x - 80 = 4 \times y.$$

Rearranging we then need to solve the system of equations

$$\begin{cases} x + y & = & 100, \\ 2x - 4y & = & 80. \end{cases}$$

Multiplying the first equation by 4,

$$4x + 4y = 400,$$

so if we add the two equations together we have

$$6x = 480$$

and thus solving for x,

$$x = 80.$$

Substituting into the first equation,

$$80 + y = 100$$

so solving for y,

$$y = 100 - 80 = 20.$$

Hence there are 80 birds and 20 cats.

Problem 5.2 Applications to Mixing Problems

(a) Connor is helping his biology teacher get ready for the class's next laboratory. The teacher needs 10 liters of a 68% alcohol solution for the students to disinfect their tools. Connor makes this solution by mixing a 80% alcohol solution with a 50% alcohol solution. How many liters of each solution did Connor use?

Answer

50% solution: 4 liters, 80% solution: 6 liters

Solution

Since Connor creates 10 liters of 68% alcohol solution, he uses a total of

$$68\% \times 10 = 6.8$$

liters of pure alcohol. If he used only the 50% solution, he would only have

$$50\% \times 10 = 5$$

liters of pure alcohol, which is

$$6.8 - 5 = 1.8$$

liters less than he actually needs. Every liter of 80% solution he uses instead of a liter of 50% solution has

$$80 - 50 = 30$$

percent more alcohol (so contains 0.3 extra liters of alcohol). Hence, Connor used

$$1.8 \div 30\% = 1.8 \div 0.3 = 6$$

liters of the 80% alcohol solution. The other

$$10 - 6 = 4$$

liters are therefore the 50% alcohol solution.

(b) Debbie is mixing orange juice concentrate for her restaurant. The first juice concentrate is 64% real orange juice. The second is only 48% real orange juice. How many ounces of 48% real orange juice should she use to make 1600 ounces of 58% real juice?

Answer

600 ounces

Solution 1

If Debbie use 1600 ounces are from the first kind of orange juice, there would be

$$64\% \times 1600 = 0.64 \times 1600 = 1024$$

ounces of orange juice. She wants the juice to be 58% real orange juice, so she actually needs

$$58\% \times 1600 = 928$$

ounces of real orange juice. Hence she needs

$$1024 - 928 = 96$$

less ounces. Each ounce of the second kind contains

$$64\% - 48\% = 16\%$$

less orange juice than the first kind. Debbie needs 96 less ounces, so she needs to replace

$$96 \div 16\% = 96 \div 0.16 = 600$$

ounces of the first kind with the second kind. Hence Debbie should use 600 ounces of the 48% real orange juice.

Solution 2

(Algebra) Assume there are x ounces from the first kind and y ounces from the second kind. There are 1600 ounces in total, so

$$x + y = 1600.$$

In total we want a 58% juice mixture, so we need

$$58\% \times 1600 = 928$$

ounces of pure orange juice. Since the pure juice comes from either the first juice or the second,

$$64\% \times x + 48\% \times y = 928.$$

This gives the system of equations

$$\begin{cases} x + y & = & 1600, \\ 0.64x + 0.48y & = & 928. \end{cases}$$

Solving the first equation for x we get

$$x = 160 - y.$$

We can then plug this in the second equation to get

$$0.64 \times (1600 - y) + 0.48y = 928,$$

after distributing we have

$$1024 - 0.64y + 0.48y = 928$$

so by combining like terms we have

$$-0.16y = -96.$$

Therefore we can solve for y to get

$$y = \frac{-96}{-0.16} = \frac{96}{0.16} = 600,$$

the number of ounces of the second kind of orange juice.

Problem 5.3 Answer the following

(a) The owner of the Fancy Food Shoppe wishes to mix cashews selling at $8.00 per kilogram and pecans selling at $7.00 per kilogram. How much of each kind of nut should be mixed to get 8 kg worth $7.25 per kilogram?

Answer

Cashews: 2 kg, Pecans: 6 kg

Solution 1

Since the owner wants the 8 kg mixture to be worth $7.25 per kilogram, in total it should cost

$$8 \times 7.25 = 58$$

dollars. If the owner gets all 8 kilograms from pecans, the mixture would cost

$$8 \times 7.00 = 56$$

dollars,

$$58 - 56 = 2$$

less than the actual cost. The extra $2 comes from cashews. Each kilogram of cashews costs

$$8.00 - 7.00 = 1$$

dollar more than a kilogram of pecans, and replacing 2 pounds of pecans with cashews will increase the price of the mixture to the correct $58. The remaining

$$8 - 2 = 6$$

kilograms are from pecans.

Solution 2

(Algebra) Assume there are x kilograms of cashews and y kilograms of pecans. We first have

$$x + y = 8$$

as the owner want 8 kg in total. The 8 kg mixture should costs $7.25 per kg, or

$$8 \times 7.25 = 58$$

dollars in total. Hence

$$8 \times x + 7 \times y = 58$$

and we have the system of equations

$$\begin{cases} x + y & = & 8, \\ 8x + 7y & = & 58. \end{cases}$$

Multiply the first equation by 7 to get

$$7x + 7x = 56$$

so if we subtract this from the second equation we have

$$x = 58 - 56 = 2.$$

Substituting back into the first equation,

$$2 + y = 8$$

so

$$y = 8 - 2 = 6.$$

Therefore the owner should use 2 kilograms of cashews and 6 kilograms of pecans.

(b) A meat distributor paid $2.50 per pound for hamburger meat and $4.50 per pound for ground sirloin. How many pounds of each did he use to make 100 pounds of meat mixture that will cost $3.24 per pound?

Answer

37 pounds sirloin, 63 pounds hamburger

Solution 1

The meat distributor wants a mixture that costs \$3.24 per pound, so 100 pounds will cost

$$100 \times 3.24 = 324$$

dollars. 100 pounds of ground sirloin costs

$$100 \times 4.50 = 450$$

dollars, which is

$$450 - 324 = 126$$

dollars more expensive than the goal. Each pound of hamburger meat is

$$4.50 - 2.50 = 2$$

dollars cheaper than a pound of ground sirloin, so replacing

$$126 \div 2 = 63$$

pounds of ground sirloin with hamburger meat will reduce the price to \$324 as needed. Hence the distributor used

$$100 - 63 = 37$$

pounds of ground sirloin and 63 pounds of hamburger meat.

Solution 2

(Algebra) Assume there are x pounds of hamburger meat and y pounds of ground sirloin. The total mixture is 100 pounds so

$$x + y = 100.$$

The meat distributor wants an entire 100 pound mixture that costs \$3.24 per pound, which costs

$$100 \times 3.24 = 324$$

dollars in total. Since this mixture is a combination of hamburger meat and ground sirloin,

$$2.50 \times x + 4.50 \times y = 324.$$

This gives the system of equations

$$\begin{cases} x + y &= 100, \\ 2.5x + 4.5y &= 324. \end{cases}$$

Solving the first equation for y,
$$y = 100 - x.$$

Substituting into the second equation,

$$2.5x + 4.5 \times (100 - x) = 324$$

so distributing we have
$$2.5x + 450 - 4.5x = 324.$$

Combining like terms we get
$$-2x = -126$$

so

$$x = \frac{-126}{-2} = 63.$$

Lastly we solve for y,
$$y = 100 - 63 = 37.$$

Hence the meat distributor should use 37 pounds of hamburger meat and 63 pounds of ground sirloin.

Problem 5.4 Solve the following questions.

(a) A candy shop sold three flavors of candies, cherry, strawberry, and watermelon, in the morning. The prices are \$20/kg, \$25/kg, and \$30/kg, respectively. The shop sold a total of 100 kg and received \$2570. It is known that the total sale of cherry and watermelon flavor candies combined is \$1970. How many kilograms of watermelon flavor candies were sold?

Answer

45

Solution 1

We know the total sale was \$2570. Since \$1970 of this was for cherry and watermelon, the remaining
$$2570 - 1970 = 600$$

dollars was due to strawberry. Hence

$$600 \div 25 = 24$$

kilograms of strawberry candy was sold. Thus the other

$$100 - 24 = 76$$

kilograms were cherry and watermelon. If thse 76 kg were all cherry, the total sales would be

$$76 \times 20 = 1520$$

dollars, which is

$$1970 - 1520 = 450$$

less than the actual amount. As each kg of watermelon is

$$30 - 20 = 10$$

dollars per kg more expensive, there must have been

$$450 \div 10 = 45$$

kg of watermelon flavor candies sold.

Solution 2

Let x be the amount of cherry candy sold in kg, y the amount of strawberry candy, and z the amount of watermelon candy. There is 100 kg sold in total, so

$$x + y + z = 100.$$

We then know cherry costs \$20 per kg, strawberry \$25 per kg, and watermelon \$30 per kg, so we have that

$$20 \times x + 25 \times y + 30 \times z = 2570,$$

the total amount sold in dollars. We also know \$1970 of this was just cherry and watermelon, so

$$20 \times x + 30 \times z = 1970.$$

This gives the system of equations

$$\begin{cases} x + y + z & = & 100, \\ 20x + 25y + 30z & = & 2570, \\ 20x + 30z & = & 1970. \end{cases}$$

Noticing the similarities between the second and third equations, we see that if we subtract the third from the second we get

$$25y = 2570 - 1970 = 600$$

so we can solve for

$$y = \frac{600}{25} = 24.$$

Plugging this into the first equation,

$$x + 24 + z = 100$$

so simplifying and multiplying by 20 we get

$$20x + 20z = 1520.$$

Subtracting this from the third equation we have

$$10z = 1970 - 1520 = 400.$$

Hence,

$$z = \frac{400}{10} = 40,$$

the number of kg of watermelon candy sold, as needed.

(b) The school purchases 3 different sizes of projectors, total of 47. The large size costs \$700, the medium costs \$300, and the small costs \$200. The total cost of the projectors is \$21200, and there are twice as many medium projectors than the small. How many large projectors does the school purchase?

Answer

20

Solution 1

If the school bought only large projectors, they would spend a total of

$$47 \times 700 = 32900$$

dollars, which is

$$32900 - 21200 = 11700$$

dollars over their budget of 21200. Since there are twice as many medium projectors as small projectors, we can view them as coming in groups of 3, with 1 small and 2 medium projectors. One such group costs a total of

$$1 \times 200 + 2 \times 300 = 800$$

dollars, which is
$$3 \times 700 - 800 = 1300$$

dollars cheaper than a group of 3 large projectors. Hence if the school switchs
$$11700 \div 1300 = 9$$

groups of 3 large projectors to groups of 1 small, 2 medium projectors the school will spend the correct amount of money. Thus, the school buys
$$47 - 3 \times 9 = 47 - 27 = 20$$

large projectors.

Solution 2

(Algebra) Let x be the number of small projectors, so we know the number of medium projectors is
$$2 \times x = 2x.$$

Let y be the number of large projectors. There are 47 projectors bought in total, so
$$x + 2x + y = 47,$$

for a total cost of \$21200, so
$$200 \times x + 300 \times (2x) + 700 \times y = 21200.$$

Simplifying these two equations we get the system of equations
$$\begin{cases} 3x + y & = & 47, \\ 800x + 700y & = & 21200. \end{cases}$$

Dividing the second equation by 100 we get a simplified version of
$$8x + 7y = 212.$$

Solving the first equation for y,
$$y = 47 - 3x.$$

Plugging this into the other equation we get
$$8x + 7 \times (47 - 3x) = 212,$$

or
$$8x + 329 - 21x = 212$$

after distributing. Hence we get

$$-13x = -117$$

after combining like terms and hence

$$x = 9.$$

Therefore,

$$y = 47 - 3 \times 9 = 20.$$

Thus the school purchases 20 large projectors.

Problem 5.5 Connor is classifying bugs for a biology project. He has 15 bugs in total. Some are spiders, with 8 legs. Some are houseflies, with 6 legs and 1 pair of wings. The remaining are dragonflies, with 6 legs and 2 pairs of wings. If there are 98 legs and 17 pairs of wings, how many of each type of bug does Connor have?

Answer

4 spiders, 5 houseflies, 6 dragonflies

Solution

If we first assume all 15 bugs are houseflies, then there will be

$$15 \times 6 = 90$$

legs. There are in fact 98 legs, which is

$$98 - 90 = 8$$

more than if we all bugs were houseflies. Each dragonfly has the same number of legs but each spider has

$$8 - 6 = 2$$

extra legs, so there must be

$$8 \div 2 = 4$$

spiders in total. The number of houseflies and dragonflies is thus

$$15 - 4 = 11.$$

Again, assume these 11 bugs are all houseflies. There will be 11 pairs of wings, which is

$$17 - 11 = 6$$

less than the actual amount. Each dragonfly has an extra pair of wings, so there must be 6 dragonflies. Hence the number of houseflies is

$$11 - 6 = 5.$$

To summarize there are 4 spiders, 5 houseflies, and 6 dragonflies.

Problem 5.6 A crab has 10 legs. A mantis has 6 legs and 1 pair of wings. A dragonfly has 6 legs and 2 pairs of wings. There are a total of 37 of the three types. There are 250 legs in total. There are 52 pairs of wings in total. How many of each kind are there?

Answer

7 crabs, 8 mantises, 22 dragonflies

Solution 1

If all 37 animals are mantises, there will be a total of

$$37 \times 6 = 222$$

legs, which is

$$250 - 222 = 28$$

legs more than the actual amount. Each dragonfly also has 6 legs, but each crab has

$$10 - 6 = 4$$

extra legs. Hence there must be

$$28 \div 4 = 7$$

crabs to give us the 28 extra legs. The remaining

$$37 - 7 = 30$$

are either mantises or dragonflies. They each have the same number of legs, so we need to look at the pairs of wings. If we again assume all 30 remaining are mantises, there will be a total of

$$30 \times 1 = 30$$

pairs of wings, which is

$$52 - 30 = 22$$

less than the actual amount. Each dragonfly has

$$2 - 1 = 1$$

extra pair of wings, so there must be 22 dragonflies in total. Lastly, the remaining

$$30 - 22 = 8$$

are mantises. In all there are 7 crabs, 8 mantises, and 22 dragonflies.

Solution 2

(Algebra) Let x be the number of crabs, and y be the number of mantises, and z be the number of dragonflies. By the tota number of animals we have

$$x + y + z = 37.$$

Using the number of legs,

$$10 \times x + 6 \times y + 6 \times z = 250.$$

Finally, using the number of pairs of wings,

$$0 \times x + 1 \times y + 2 \times z = 52.$$

This gives us the system of equations

$$\begin{cases} x + y + z & = & 37, \\ 10x + 6y + 6z & = & 250, \\ y + 2z & = & 52. \end{cases}$$

Multiplying the first equation by 6,

$$6x + 6y + 6z = 222,$$

so after subtracting this from the second equation we have

$$4x = 28$$

and therefore,

$$x = \frac{28}{4} = 7.$$

Plugging this into the second equation,

$$10 \times 7 + 6y + 6z = 250$$

so

$$6y + 6z = 180.$$

Dividing this by 6 we have

$$y + z = 30.$$

Subtracting this from the third equation,

$$z = 22,$$

and therefore

$$y + 22 = 30$$

so

$$y = 30 - 22 = 8.$$

Hence, $x = 7$, $y = 8$, and $z = 22$, so there are 7 crabs, 8 mantises, and 22 dragonflies.

Problem 5.7 A turtle has 4 legs and a crane has 2 legs. There are totally 100 heads of turtles and cranes, and there are 20 more crane legs than turtle legs. How many of each animal are there?

Answer

30 turtles, 70 cranes

Solution 1

There are 20 more crane legs than turtle legs. Each turtle has 4 legs, so if we assume

$$20 \div 4 = 5$$

additional turtles are added, then there are

$$100 + 5 = 105$$

heads altogether and the number of crane legs and turtle legs are the same. Since each turtle has twice as many legs as a crane, there must be half as many turtles. Since

$$105 \div 3 = 35$$

there are 35 turtles and

$$2 \times 35 = 70$$

cranes now. Removing the added 5 turtles, the original numbers are

$$35 - 5 = 30$$

turtles and 70 cranes.

Solution 2

(Algebra) Let x be the number of turtles and y the number of cranes. There are 100 heads in total, so

$$x + y = 100.$$

We know there are 20 more crane legs than turtle legs. Hence

$$4x + 20 = 2y.$$

After combining like terms we get the system of equations

$$\begin{cases} x + y & = & 100, \\ 4x - 2y & = & -20. \end{cases}$$

Multiplying the first equation by 2 we get

$$2x + 2y = 200,$$

and then adding this to the second equation we have

$$6x = 180.$$

Hence we can solve for x to get

$$x = \frac{180}{6} = 30$$

as the number of turtles. Substituting into the first equation in our system,

$$30 + y = 100$$

so

$$y = 100 - 30 = 70,$$

the number of cranes.

Problem 5.8 A party store has 100 balloons coming in either large or small sizes. The large balloons have a volume of 5 liters and are filled with 80% helium and 20% regular air. The small balloons have a volume of 2 liters and are filled with 90% helium and 10% regular air. If 312 liters of pure helium was used to fill the balloons, how many liters of regular air were used to fill the balloons?

Answer

68

Solution

First note that a large balloon contains $5 \times 80\% = 4$ liters of pure helium and a small balloon contains $2 \times 90\% = 1.8$ liters of pure helium.

If all 100 balloons were large, then there would be a total of

$$100 \times 4 = 400$$

liters of pure helium used, which is $400 - 312 = 88$ liters extra. Each large baloon contains $4 - 1.8 = 2.2$ extra liters of pure helium, so we have to change

$$88 \div 2.2 = 40$$

of the balloons to small balloons. Hence there are 40 small balloons and 60 large balloons.

Hence the total volume of all 100 balloons is

$$40 \times 2 + 60 \times 5 = 80 + 300 = 380,$$

and therefore

$$380 - 312 = 68$$

liters of regular air were used to fill the balloons.

Problem 5.9 Some chickens and rabbits have a total of 100 feet. If each chicken was exchanged for a rabbit, and each rabbit was exchanged for a chicken, there would be a total of 86 feet. How many chickens are there? How many rabbits?

Answer

12 chickens, 19 rabbits

Solution 1

We know that there are 100 feet total. We first find out how many animals there are in total. Pretend that for every chicken on the farm, we pair it up with a new rabbit, and for every rabbit on the farm, we pair it up with a new chicken. Note that this adds a total of 86 feet. Hence there are a total of

$$100 + 86 = 186$$

feet. Each chicken and rabbit pair has a combined total of

$$2 + 4 = 6$$

feet, so there must be

$$186 \div 6 = 31$$

chicken and rabbit pairs. As every original animal is in exactly one pair, this means there are 31 animals on the farm.

If all 31 animals were chickens, there would be a total of

$$31 \times 2 = 62$$

feet, which is

$$100 - 62 = 38$$

less than the true amount. As each rabbit has

$$4 - 2 = 2$$

extra feet, if we change

$$38 \div 2 = 19$$

chickens to rabbits we will have the correct number of feet. Hence there are

$$31 - 19 = 12$$

chickens and 19 rabbits.

Solution 2

(Algebra) Let x be the number of chickens and y be the number of rabbits. There are 100 feet in total, so as each chicken has 2 feet and each rabbit has 4,

$$2 \times x + 4 \times y = 100.$$

We also know that swapping all the animals we have 86 feet, so

$$4 \times x + 2 \times y = 86.$$

This gives the system of equations

$$\begin{cases} 2x + 4y &=& 100, \\ 4x + 2y &=& 86. \end{cases}$$

Doubling the first equation gives us

$$4x + 8y = 200.$$

We can then subtract the second equation from this to get

$$6y = 114,$$

and dividing by 6 we have

$$y = \frac{114}{6} = 19.$$

Substituting into the first equation,

$$2x + 4 \times 19 = 100$$

so combining like terms we have

$$2x = 24.$$

Hence

$$x = \frac{24}{2} = 12$$

so there are 12 chickens and 19 rabbits.

Problem 5.10 Bella goes shopping at the marketplace for shawls and belts. The shawls she likes each cost \$12. The belts she likes each cost \$14. Bella has exactly enough money to buy a certain number of shawls. If she buys belts instead, she has exactly enough money to buy 3 fewer belts. How much money did Bella bring with her to the market?

Answer

\$252

Solution 1

Bella can buy 3 more shawls than belts. These 3 shawls cost a total of

$$3 \times 12 = 36$$

dollars in total. Every shawl is

$$14 - 12 = 2$$

dollars cheaper than a belt. Hence if Bella buys

$$36 \div 2 = 18$$

shawls instead of 18 belts, she will have enough leftover money to buy 3 extra shawls. Hence Bella has exactly enough money to buy 18 belts, which is

$$18 \times 14 = 252$$

in total.

Solution 2

(Algebra) Let x be the number of belts Bella can buy. Then the number of shawls Bella can afford is $x + 3$. Buying x belts costs Bella

$$14 \times x$$

dollars, while buying $x + 3$ shawls costs

$$12 \times (x + 3).$$

Since Bella would spend all her money with either purchase, the two expressions are equal:

$$14x = 12 \times (x + 3).$$

Distributing and combining like terms we have

$$2x = 36$$

so dividing by 2 we have

$$x = \frac{36}{2} = 18.$$

Thus Bella has enough money to buy 18 belts. Hence she has

$$18 \times 14 = 252$$

dollars in total.

6 Solutions to Chapter 6 Examples

Problem 6.1 Introductory Motion Questions

(a) Samantha needs to run a mile for less than 10 minutes for her PE class. What is the minimum average speed she must run to complete the mile in time? Give your answer in both miles per hour and meters per second (use the approximation 1 mile is roughly 1600 meters).

Answer

6 miles per hour, ≈ 2.7 meters per second

Solution

There are 60 minutes in one hour. If Samantha has to run 1 mile in 10 minutes, then she will have to run 6 miles in one hour, her least average speed is 6 miles per hour.

Since 1 mile is roughly 1600 meters and 1 hours is 3600 seconds, we have

$$6 \text{ mph} = 6 \times 1600 \div 3600 = \frac{1600}{600} = \frac{8}{3} \approx 2.66 \text{ m/s}.$$

(b) Jimmy and his brother took a circular ride at an amusement park on averages of 30 miles per hour and took them $2\frac{1}{2}$ minutes. Roughly how big is the diameter of the circular track?

Answer

≈ 0.398 miles

Solution

In order to find the distance they traveled, all we need to do is to multiply the speed and the amount of time they used. However, the unit of the speed is miles per hour, while the unit of the time is minutes. Now, we need to convert one of the unit, so that we can multiple them.

$$2.5 \text{ minutes } = 2.5 \times \frac{1}{60} = \frac{1}{24} \text{ hours}.$$

Now we can find the distance,

$$30 \times \frac{1}{24} = 1.25 \text{ miles.}$$

We are asked to find the rough measurement of the diameter of the track. Let's assume the track is a perfect circle. The formula for perimeter of the circle is the diameter multiplied by a constant π. So the diameter $= \dfrac{1.25}{\pi} \approx 0.398$ miles.

Problem 6.2 The Winchers family is taking a road trip from Los Angeles, California to Phoenix, Arizona. The distance is about 360 miles.

(a) If they drove at a constant speed of 60 miles per hour for the first 150 miles, and drove at a constant speed of 70 miles per hour for the remaining 210 miles, how long did it take them to get there?

Answer

5.5 hours

Solution

In the first 150 miles, the speed is 50 miles per hour, thus it took them $150 \div 60 = 2.5$ hours. For the remaining distance 210 miles, they drove 70 miles per hour, so it took them $210 \div 70 = 3$ hours. The total time is $2.5 + 3 = 5.5$ hours.

(b) Suppose instead the family plans to get there in 6 hours. If they drove at a constant speed of 50 miles per hour for the first 120 miles, at what constant speed do they need to drive for the remaining 240 miles in order to get there in time?

Answer

≈ 66.7 mph

Solution

In the first 120 miles, the speed is 50 miles per hour, thus it took $120 \div 50 = 2.4$ hours. In order to get there in 6 hours, they need to drive the remaining 240 miles in $6 - 2.4 = 3.6$ hours, and the speed would be

$$240 \div 3.6 = 200 \div 3 \approx 66.7$$

miles per hour.

Problem 6.3 Relative Speeds

(a) John's house and Mary's house are 14 miles apart. They start at noon to walk toward each other in order to go to a book fair together. John walks at a rate of 3 mph, and Mary walks at a rate of 4 mph. How many hours will it take them to meet?

Answer

2 hours

Solution 1

(Use Relative Speed) John and Mary are traveling in opposite directions toward each other, therefore, their relative speed is the sum of their two speeds, which is $3 + 4 = 7$ miles per hour towards each other. Therefore, it takes them $14 \div 7 = 2$ hours until they meet.

Solution 2

(Algebra) Let the time it takes for them to meet each other be x hours. In x hours, John travels
$$3 \times x$$
and Mary travels
$$4 \times x$$
miles, so if they meet up at that time,
$$3x + 4x = 14.$$
Solve for x, we have
$$x = 2,$$
so it takes 2 hours for them to meet each other.

(b) Terry and Susan are entered in a 24-mile race. Susan's average rate is 4 miles per hour and Terry's average rate is 6 miles per hour. Both start at the same time. How far will Susan be away from the finish line when Terry crosses the line?

Answer

8 miles

Solution

First find the time that takes Terry to finish the race, which is

$$24 \div 6 = 4 \text{ hours.}$$

After 4 hours of running, the distance that Susan has run is

$$4 \times 4 = 16 \text{ miles.}$$

Since the race is 24 miles long, Susan still has

$$24 - 16 = 8 \text{ miles}$$

until she reaches the finish line.

Problem 6.4 Round Trips

(a) Mr. Winchers drives a car from home to a client site in Wilmington early morning at the speed of 72 miles per hour. On his way back home from the client site in Wilmington, the traffic is getting bad; he drives the car back home at a reduced speed of 48 miles per hour. What is his average speed for the round trip?

Answer

57.6 mph

Solution

We are not given a distance in the problem so let's assume the distance each way is 144 miles (since 144 is a multiple of both 72 and 48). Therefore, on the way there the drive takes $144 \div 72 = 2$ hours while on the return trip it takes $144 \div 48 = 3$ hours. Hence the total trip takes $2 + 3 = 5$ hours. As the total trip is $144 + 144 = 288$ miles, the average speed is $288 \div 5 = 57.6$ miles per hour.

Note double check that you get the same answer using different distances! You can also let d denote the one way distance and see that the final answer will not depend on d.

(b) Mike drives his car for a round trip between LA and San Diego. He drives at 70 miles per hour to get from LA to San Diego. At what speed should he drive back, if his average speed for the round trip is 60 miles per hour?

Answer

$\dfrac{105}{2}$ miles per hour

Solution

We are not given any distances, so assume the distance between LA and San Diego is 210 miles, so the round trip distance is 420 miles. If the average speed for the round trip is 60 miles per hour, the round trip will take

$$420 \div 60 = 7 \text{ hours.}$$

hours. Then, we can find the time it took Mike to travel from LA to San Diego, which is

$$210 \div 70 = 3 \text{ hours.}$$

So the time he used to drive back is

$$7 - 3 = 4 \text{ hours.}$$

Therefore, the speed he must drive back is

$$210 \div 4 = \frac{105}{2} \text{ miles per hours.}$$

Problem 6.5 One day, Bob rode his bike to school. When school is off, he forgot his bike and walked home instead. He spent a total of 50 minutes on the road for the round trip. If he walked for both directions, he would have spent a total of 70 minutes. How much would be the total time if he rode his bike for both directions?

Answer

30 minutes

Solution

If it takes Bob 70 minutes to walk to and from school, his walk home from school took

$$70 \div 2 = 35$$

minutes. Therefore, the time it took for him to ride the bike to school was

$$50 - 35 = 15$$

minutes. If Bob instead had biked both ways, it would take

$$15 \times 2 = 30$$

minutes to travel to and from school.

Problem 6.6 Ling goes mountain hiking in a park. She first walks uphill at a speed of 2.5 miles per hour, and she next walks downhill at a speed of 4 miles per hour. The round trip takes 3.9 hours. What is the distance for the round trip?

Answer

12 miles

Solution 1

Since Ling walks the same distance up and down the mountain, let's assume the distance is 10 miles. So the time it take Ling to reach the top of the mountain is

$$10 \div 2.5 = 4 \text{ hours},$$

and the walk back downhill takes

$$10 \div 4 = 2.5 \text{ hours}.$$

The round trip therefore takes

$$4 + 2.5 = 6.5 \text{ hours}.$$

To find the average speed, we calculate the total distance divided by the total time. The total distance is 20 miles and the total time is 6.5 hours, so the average speed is $20 \div 6.5 = \dfrac{20}{6.5}$ miles per hour. Since the round trip takes 3.9 hours, we can find the distance $\dfrac{20}{6.5} \times 3.9 = 12$ miles for the round trip.

Solution 2

Let x be the distance from the bottom to the top of the mountain. The time it take Ling to reach the top of the mountain is

$$x \div 2.5 = \frac{2x}{5} \text{ hours},$$

and the walk back downhill takes

$$x \div 4 = \frac{x}{4} \text{ hours.}$$

The round trip therefore takes

$$\frac{2x}{5} + \frac{x}{4} = \frac{13x}{20} \text{ hours.}$$

Therefore

$$\frac{13x}{20} = 3.9$$

so solving for x we get

$$x = 6.$$

Since this is the distance one way, the round trip is 12 miles.

Problem 6.7 At 6 AM, bus station A starts to dispatch buses to station B, and station B starts to dispatch buses to station A. They each dispatch one bus to the other station every 8 minutes. The one-way trip takes 45 minutes. One passenger gets on the bus at station A at 6:16 AM. How many buses coming from station B will the passenger see en route?

Answer

8

Solution

The passenger gets on the bus at 6:16 AM to travel to station B. Since the one-way trip takes 45 minutes, none of the buses from station B have arrived at station A when the passenger departs. The passenger will therefore see every bus that departs station B between 6 AM and their arrival time at station B which is 7:01 AM. One bus leaves exactly at 6 AM, and in the 61 minutes that follow 7 more buses leave, because

$$61 \div 8 \approx 7.625$$

and we round down because the buses leave at the end of each 8 minute interval. Therefore, the passenger sees a total of

$$7 + 1 = 8$$

buses en route from station A to station B.

Problem 6.8 Non-Constant Motion

(a) Sam walks up a hill. After every 30 minutes of walking he takes 10 minutes to rest. When he walks down the hill, he instead rests for 5 minutes after every 30 minutes of walking. Sam walks downhill 1.5 times faster than he walks uphill. If he spends 3 hours and 50 minutes traveling up the hill, how much time does he spend traveling down the hill?

Answer

2 hours 15 minutes

Solution

We know it takes 3 hours and 50 minutes, or

$$3 \times 60 + 50 = 230$$

minutes to travel uphill. We want to find how much of this time Sam was actually walking. A walking/resting cycles takes 40 minutes, so because

$$230 = 5 \times 40 + 30$$

the
final 30 minutes. Hence, he walks a total of

$$5 \times 30 + 30 = 180$$

minutes. Since Sam walks downhill at a speed 1.5 times faster as that he walks uphill, he take will spend 1.5 times less time walking downhill, a total of

$$180 \div 1.5 = 120$$

minutes. In these 120 minutes, me must rest

$$(120 \div 30) - 1 = 3$$

times (since he can walk the final 30 minutes). He therefore spends a total of

$$120 + 3 \times 5 = 135$$

minutes, or 2 hours and 15 minutes to travel uphill.

(b) A hunting dog chases a hare 21 meters ahead. The dog runs in a series of jumps, with each jump being 3 meters long. Each jump for the hare is 2.1 meters. If the dog jumps three times for every four times the hare jumps, how much farther can the hare travel before the dog catches it?

Answer

294 meters

Solution 1

Each jump for the dog is 3 meters, so in 3 jumps the dog moves a total of

$$3 \times 3 = 9$$

meters. In the same time the hare jumps 4 times, for a total of

$$4 \times 2.1 = 8.4$$

meters. Therefore, every time the hare jumps 4 times, the dog catches up

$$9 - 8.4 = 0.6$$

meters. Since

$$21 \div 0.6 = 21 \div 3/5 = 35$$

the hare can jump a total of

$$35 \times 4 = 140$$

times before the dog catches up. Thus, the hare travels a total of

$$140 \times 2.1 = 294$$

meters before the dog catches it.

Solution 2

(Algebra) Since the dog jumps three times for every four times the hare jumps, the ratio of jumps is $3 : 4$. Let x be such that the dog jumps $3x$ times and the hare jumps $4x$ times when the dog catches the hare. In that time, the dog will travel

$$3 \times 3x = 9x$$

meters. The hare starts 21 meters ahead and jumps a total of

$$2.1 \times 4x = 8.4x$$

meters. Since the dog catches up to the hare,

$$9x = 20 + 8.4x$$

Combining like terms we have

$$0.6x = 21$$

so we can solve for x, to get

$$x = 21 \div 0.6 = 21 \div \frac{3}{5} = 35.$$

Therefore the hare can travel

$$2.1 \times 4 \times 35 = 294$$

more meters before the dog catches it.

Problem 6.9 Answer the following questions.

(a) Cindy rides her bike from home to school at a speed that is 120 meters per minute faster than if she walks, and the time she spends is 3/5 less than if she walks. How fast does Cindy walk from home to school?

Answer

80 meters per minute

Solution 1

We are not given a distance or time in the problem, so we may assume for convenience that Cindy walks to school for 5 minutes. Since the time she spends biking is $\frac{3}{5}$ less than the time she spends walking, she spends

$$5 \times \frac{2}{5} = 2$$

minutes biking to school. Since Cindy bikes 120 meters per minute faster than she walks (and can bike to school in 2 minutes), if Cindy walks for 2 minutes, she will have

$$2 \times 120 = 240$$

meters left to walk before she gets to school. Since it takes her 5 minutes in total to talk to school, she can walk 240 meters in

$$5 - 2 = 3$$

minutes. Hence, Cindy can walk

$$240 \div 3 = 80$$

meters per minute.

Solution 2

(Algebra) Let the speed that Cindy walks to be x meters per minute. Since the time she spends biking is $\frac{3}{5}$ less than the time she spends walking, she bikes for $\frac{2}{5}$ of the time she bikes. Since the distances she travels are the same, we have

$$x = \frac{2}{5} \times (x + 120).$$

Distributing and grouping like terms we have

$$\frac{3}{5}x = 48.$$

Then we can solve for x,

$$x = 48 \times \frac{5}{3} = 80.$$

Therefore, Cindy walks at a speed of 80 meters per minute.

(b) Joe and JoAnn walk toward each other from two locations that are 36 miles apart. If Joe departed 2 hours earlier, they would meet 2.5 hours after JoAnn departed. If JoAnn departed 2 hours earlier, they would meet 3 hours after Joe departed. Find the respective speed at which each walks.

Answer

Joe: 6 miles/hr, JoAnn: 3.6 miles/hr

Solution 1

Since the two locations are 36 miles apart, when Joe and JoAnn meet they have walked a combined 36 miles. In the first scenario, Joe walks a total of 4.5 hours and JoAnn walks a total of 2.5 hours. In the second scenario, Joe walks a total of 3 hours and JoAnn walks a total of 5 hours. If we double the first scenario, we see that if Joe walks

$$4.5 \times 2 = 9$$

hours and JoAnn walks for

$$2.5 \times 2 = 5$$

hours, they walk a combined

$$36 \times 2 = 72$$

miles. Comparing this with the second scenario (since JoAnn walks 5 hours in each case), we see that in

$$9 - 3 = 6$$

hours, Joe must walk

$$72 - 36 = 36$$

miles by himself. Therefore Joe walks

$$36 \div 6 = 6$$

miles/hr. Thus, in 3 hours, Joe walks

$$6 \times 3 = 18$$

miles, so using the second scenario JoAnn must walk the remaining

$$36 - 18 = 18$$

miles in 5 hours. Thus, JoAnn walks at a speed of

$$18 \div 5 = 3.6$$

miles/hr.

Solution 2

Let's assume the Joe's walking speed is x miles/hr, and JoAnn's walking speed is y miles/hr. If they meet 2.5 hours after JoAnn departs, JoAnn walks for a total of 2.5 hours, while Joe walks 2 extra hours, for a total of 4.5 hours. This gives us the equation

$$4.5 \times x + 2.5 \times y = 36.$$

If they meet 3 hours after Joe departs, Joe walks a total of 3 hours. JoAnn walks 2 extra hours, so she walks 5 hours. This gives another equation

$$3 \times x + 5 \times y = 36.$$

This gives us the system of equations

$$\begin{cases} 4.5x + 2.5y &= 36, \\ 3x + 5y &= 36. \end{cases}$$

Multiplying the first equation by 2 we get

$$9x + 5y = 72.$$

If we subtract the second equation from this we have

$$6x = 36,$$

so

$$x = 6.$$

Substituting back into the second equation we have

$$3 \times 6 + 5y = 36$$

so combining like terms we have

$$5y = 18$$

and solving for y gives

$$y = \frac{18}{5} = 3.6$$

Therefore, Joe walks at a speed of 6 miles/hr, and JoAnn walks at a speed of 3.6 miles/hr.

Problem 6.10 Two trains are 121 meters and 99 meters in length respectively. They travel at a rate of 40 km/h and the other at the rate of 32 km/h.

(a) If the trains are moving in opposite directions, how long would it take them to completely clear each other from the moment they meet?

Answer

11 seconds

Solution

Since the two trains are moving in opposite directions, then the relative speed is the sum of the two speeds,

$$40 + 32 = 72 \text{ km/h}$$

In order for the two trains to be completely clear of each other from the moment they meet, the total distance the two trains traveled is the sum of the length of the two trains, which is

$$121 + 99 = 220 \text{ meters.}$$

Notice that the units for the speed and the length of the train are not consistent. Hence, we need to convert one to the other before we go forward. One way is to change the speed from km/h to m/s. We know that 1 km = 1000 m and 1 hour = 3600 seconds, so 72 km/h is

$$72 \times 1000 \div 3600 = 20 \text{ m/s.}$$

Now we can find the time it takes the trains to complete clear each other:

$$220 \div 20 = 11 \text{ seconds.}$$

(b) If the two trains are traveling in the same direction, how long would it take them to completely clear each other, if the faster train has just met up with the back of the slower train?

Answer

99 seconds

Solution

The two trains are moving in the same direction, so the relative speed is the difference of the two speeds, which is

$$40 - 32 = 8$$

km/h. Changing this speed into m/s (using 1 km = 1000 m and 1 hr = 3600 s),

$$8 \times 1000 \div 3600 = \frac{20}{9}$$

m/s. In order to completely clear each other, the total distance is the sum of the length of the two trains, which is

$$121 + 99 = 220$$

meters. We can then find the time it takes to do so, which is

$$220 \div \frac{20}{9} = 99$$

seconds.

7 Solutions to Chapter 7 Examples

Problem 7.1 Motion Warmups and Review

(a) A tennis ball is thrown a distance of 20 meters. What is its speed if it takes 0.5 seconds to cover the distance?

Answer

40 m/s

Solution

To find the average speed, we need to use the total distance divided by the total time, then

$$20 \div 0.5 = 40$$

gives us the average speed in m/s.

(b) It takes 6 hours for an airplane to fly a round trip. If the speed of the airplane is 1500 km per hour on the departure trip, and 1200 km per hour on the return trip. What is the one-way distance ?

Answer

4000 km

Solution 1

(Ratios) The ratio of the speeds of the airplane on the departure and return trips is

$$1500 : 1200 = 5 : 4.$$

Since the distance is the same both ways, the time spend on the trips will be in the opposite ratio $4 : 5$. Since the entire trip is 6 hours,

$$\frac{4}{9} \times 6 = \frac{8}{3}$$

hours will be spend on the departure trip. Therefore, the distance one-way is

$$1500 \times \frac{8}{3} = 4000$$

km.

Solution 2

(Algebra) Let the one-way distance to be x km. Therefore, the departure trip takes

$$x \div 1500 = \frac{x}{1500}$$

hours, while the return trip takes

$$x \div 1200 = \frac{x}{1200}$$

hours. Since the round trip takes 6 hours, we have

$$\frac{x}{1500} + \frac{x}{1200} = 6.$$

We can then solve for x to get

$$x = 4000.$$

Therefore, the distance one-way is 4000 km.

Problem 7.2 Connor and his friend Derek went to river tubing at the Math Zoom Summer Camp. Connor sat still on the tube and enjoyed the beautiful scenes on the river bank, with no effort made to pilot the tube. Derek, however, paddled down the river. Both left the starting point at 3 pm and traveled a total distance of two and a quarter miles.

(a) Connor arrived at 4:15 pm, how fast was the water flowing in the river? Use miles per hour for the speed.

Answer

1.8

Solution

Since Connor was sitting on the tube effortless, so his speed would be same as that of the water flow. The total distance is $2\frac{1}{4}$ miles and the the total time is 75 minutes, which is $1\frac{15}{60}$ hours.

To find the average speed, we calculate the total distance divided by the total time:

$$2\frac{1}{4} \div 1\frac{15}{60} = \frac{9}{4} \div \frac{75}{60} = \frac{9}{4} \times \frac{60}{75} = \frac{9}{4} \times \frac{4}{5} = \frac{9}{5} = 1.8$$

in miles per hour.

(b) Derek ended up arriving at the ending point at 3:45 pm. How fast can Derek paddle in still water?

Answer

1.2

Solution

Derek's overall speed can also be calculated with the basic motion formula by dividing the total distance by the total time. The total distance is $2\frac{1}{4}$ miles. His total time was $\frac{45}{60}$ hours.

Hence Derek's total speed was
$$2\frac{1}{4} \div \frac{45}{60} = \frac{9}{4} \div \frac{45}{60} = \frac{9}{4} \times \frac{60}{45} = \frac{9}{4} \times \frac{4}{3} = \frac{9}{3} = 3$$
Since the water speed is 1.8 miles per hours, Derek's paddling speed in still water would be the difference between his overall speed and the water flow speed. Thus Derek's paddling speed $3 - 1.8 = 1.2$ miles per hour.

Problem 7.3 Answer the following.

(a) It takes a ship 6 hours to travel downstream between two piers, and 8 hours upstream. If the water flows at a speed of 2.5 miles per hour, at what speed would the ship travel in still water?

Answer

17.5 miles per hour

Solution

Consider the same ship traveling upstream and downstream. The flow of the water adds 2.5 miles per hour upstream and subtracts 2.5 miles per hour downstream. Therefore the difference in speeds between the boat traveling upstream and downstream is 5 miles per hour. Note that in 24 hours of traveling both upstream and downstream, the ship can make
$$24 \div 6 = 4$$
trips between the piers traveling downstream and
$$24 \div 8 = 3$$

---

trips between the piers traveling upstream. Since the ship is traveling the same amount of time in both directions, the difference in the distance it travels must be due to the 5 mile per hour difference in speed due to the water. Hence, the difference between the two piers is

$$5 \times 24 = 120$$

miles. Since it takes 6 hours for the to travel downstream between the piers, it travels at a speed of

$$120 \div 6 = 15$$

miles per hour. Therefore, in still water it travels at a speed of

$$15 + 2.5 = 17.5$$

miles per hour.

(b) Mr. and Mrs. Winchers were rowing a boat upstream in the river. At exactly 1 pm, Mrs. Winchers's hat fell into the water, but they didn't find out until sometime later. They immediately turned around and rowed downstream, and caught up with the hat 6 miles from the point the hat fell. Assuming the water flow speed is 4 miles per hour. (1) What is the exact time when they found the hat? (2) Assuming they rowed at a constant speed relative to the water, what was the time when they realized the hat fell and turned around?

Answer

(1) 2:30 pm, (2) 1:45 pm

Solution

At first glance, there seem to be missing information. We don't know their rowing speed. However, since the hat was floating with the river, it took 1.5 hours for the hat to move 6 miles downstream, and that was how long it took for them to find the hat. Therefore they found the hat at 2:30pm.

For the second question, we use the hat as the frame of reference, then the water was still and the two people rowed the boat at a constant speed, so they took the same amount of time away and back. Therefore they turned around at exactly 1:45pm.

Problem 7.4 Linda goes mountain biking in a park. She first bikes on flat road at a speed of 12 miles per hour, then goes uphill at a speed of 9 miles per hour, and she next bikes downhill at a speed of 18 miles per hour, along the same trail as the uphill trip.

Finally she goes back home along the same flat road she traveled earlier. The round trip takes 4 hours. What's the distance for the round trip?

Answer

48 miles

Solution

Although we don't know the distance of each segment of the trip, we do know that the uphill and downhill distances are the same, so we can calculate the average speed of the uphill and downhill segments. For easier calculation, assume the uphill distance is 18 miles (it would give us the same average speed if we chose any other value), so the downhill distance is also 18 miles. Thus the uphill time is $18 \div 9 = 2$ hours and the downhill time is $18 \div 18 = 1$ hour. So the average speed of the uphill and downhill portion is $(18 + 18) \div (2 + 1) = 12$ miles per hour. This coincides with the flat road speed. Therefore, the average speed of the whole trip is also 12 miles per hour. Since it takes 4 hours to complete the round trip, the total distance is $12 \times 4 = 48$ miles.

Note: in this problem, the average speed of the uphill and downhill trip coincides with the flat road speed. If they didn't coincide, this solution would not work, and there would not be enough information for a definite answer.

Problem 7.5 Starting at the same time, Heather and Brenda drive their cars from site A toward site B. Heather drives at 52 km per hour, and Brenda drives at 40 km per hour. After 6 hours of driving, Heather passes a truck traveling in the opposite direction. One hour later, Brenda passes the same truck still traveling in the opposite direction. At what speed does the truck travel?

Answer

32 km/h

Solution

Since Heather and Brenda are diving in the same direction, their relative speed is the difference of their two speeds, which is

$$52 - 40 = 12$$

km/h. Therefore, after 6 hours of driving, the distance between Heather and Brenda is

$$12 \times 6 = 72$$

km. In the hour before she passes the truck, Brenda travels 40 km. Therefore, the truck must have traveled

$$72 - 40 = 32$$

km in the same hour. Thus the truck driver drives at a speed of 32 km/h.

Problem 7.6 Brian and David run along a circular track, starting from the same point, going opposite directions. They meet after 36 seconds. Assume that David runs the whole circle in 90 seconds. How long does it take Brian to run the whole circle?

Answer

60 seconds

Solution 1

Since David runs the whole circle in 90 seconds, at the time they meet (36 seconds), David finishes $\frac{36}{90} = \frac{2}{5}$ of a whole circle, so Brian finishes $1 - \frac{2}{5} = \frac{3}{5}$ of a whole circle. Thus the total time it takes for Brian to run the whole circle is $36 \div \frac{3}{5} = 60$ seconds.

Solution 2

(Algebra) Assume it takes Brian x seconds to run the whole circle, so he runs $\frac{1}{x}$ of a whole circle per second. Also we know that David runs $\frac{1}{90}$ of a whole circle per second. They meet after 36 seconds, so

$$\frac{1}{x} + \frac{1}{90} = \frac{1}{36},$$
$$\frac{1}{x} = \frac{1}{36} - \frac{1}{90} = \frac{1}{60},$$

therefore $x = 60$ seconds.

Problem 7.7 Starting at the same time, Cathy and David drive two cars toward each other from the two ends, call them A and B, of the same road. Cathy drives 1.2 times faster than David. When they pass by each other, they are 8 miles away from the halfway point between A and B. Find the total length of the road.

Answer

176 miles

Solution 1

Cathy drives 1.2 times faster than David, so since

$$1.2 = \frac{6}{5}$$

the ratio of their speeds is

$$6 : 5.$$

Since they both travel the same amount of time, this means the ratio between the distance Cathy travels and the distance David travels is also 6 : 5. We know that when they pass each other they are 8 miles from the halfway point, meaning that Cathy must have driven 16 miles more than David. Multiplying both sides of the ratio 6 : 5 by 16 we see that

$$6 : 5 = 96 : 80,$$

so Cathy must drive 96 miles and David must drive 80 miles. Therefore the total length of the road (which combined they have driven when they meet) is

$$96 + 80 = 176$$

miles.

Solution 2

(Algebra) Cathy drives 1.2 times faster than David, so since

$$1.2 = \frac{6}{5}$$

the ratio of their speeds is

$$6 : 5.$$

Since Cathy and David both travel the same about of time, the ratio between their distances is also 6 : 5. Let x be such that Cathy travels $6x$ miles and David travels $5x$ miles. They pass each other 8 miles past the halfway's point, meaning that Cathy drives 16 miles more than David. Thus,

$$6x = 5x + 16$$

so combining like terms we have

$$x = 16$$

miles. Therefore the length of the road is

$$6x + 5x = 11x = 11 \times 16 = 176$$

miles.

Problem 7.8 A railroad bridge measures 1000 meters long. A train passes the bridge. It takes 120 seconds from the time the train enters the bridge to the time the whole train gets off the bridge. There are 80 seconds during which time the whole train is on the bridge. Find both the speed and the length of the train.

Answer

Speed: 10 meters/sec. Length: 200 meters

Solution 1

Since the whole train is on the bridge for 80 seconds, it takes

$$120 - 80 = 40 \text{ seconds}$$

for the train to move on and off the bridge. Since the train takes the same amount of time to move on the bridge and move off the bridge it takes

$$40 \div 2 = 20 \text{ seconds}$$

to do each. If we focus on the head of the train, it takes

$$80 + 20 = 100 \text{ seconds}$$

for the head of the train to travel the 1000 meter length of the bridge. Hence the train moves at a speed of

$$1000 \div 100 = 10 \text{ meters/sec.}$$

Finally, since it takes 20 seconds to move off the bridge we know the length of the train is

$$20 \times 10 = 200 \text{ meters.}$$

Solution 2

(Algebra) Let the length of the train be x meters. The distance the train travels in the 120 seconds from entering the bridge to fully leaving the bridge is $1000 + x$ so the trains speed is

$$(1000 + x) \div 120 = \frac{1000 + x}{120}$$

meters/sec. However, we also know the whole train is on the track for 80 seconds, in which it travels $1000 - x$ meters, so the speed is also

$$(1000 - x) \div 80 = \frac{1000 - x}{80}$$

meters/sec. Therefore, we can set up the following equation,

$$\frac{1000+x}{120} = \frac{1000-x}{80}.$$

Clearing denominators we get

$$2000 + 2x = 3000 - 3x$$

and grouping like terms gives us

$$5x = 1000$$

so we can solve for x:

$$x = 1000 \div 5 = 200.$$

Then the speed of the train is

$$\frac{1000+200}{120} = 10$$

meters/sec. Therefore, the speed of the train is 10 m/s and the length of the train is 200 meters.

Problem 7.9 A road consists of uphill, flat and downhill sections with that order. The distances of the three sections are in ratio 1 : 2 : 3 with a total distance of 20 miles. The times JoAnn spends on the three sections are in ratio 4 : 5 : 6. She walks at a speed of 2.5 miles per hour uphill. What is the total time she spends on the road?

Answer

5 hours

Solution

We know the ratios of the three sections and the total distance, so we can find how long each section of the road is. The uphill section is

$$20 \times \frac{1}{6} = \frac{10}{3}$$

miles long, the flat section

$$20 \times \frac{2}{6} = \frac{20}{3}$$

miles long, and the downhill section

$$20 \times \frac{3}{6} = 10$$

miles long. Since JoAnn walks 2.5 miles per hour during the uphill section, it takes her

$$\frac{10}{3} \div 2.5 = \frac{4}{3}$$

hours to do so. Since the times JoAnn spends on each section are in ratio $4 : 5 : 6$ she spends

$$\frac{4}{3} \times \frac{5}{4} = \frac{5}{3}$$

hours on the flat section and

$$\frac{4}{3} \times \frac{6}{4} = 2$$

hours on the downhill section. Therefore, the total time she spends on the road is

$$\frac{4}{3} + \frac{5}{3} + 2 = 5$$

hours.

Problem 7.10 Alice, Bob, and Cindy drive their cars separately from site A to site B simultaneously. Alice drives at 60 mph and Bob drives at 48 mph. Alice passes a car from the opposite direction after 6 hours of driving. One hour later, Bob pass the same car still traveling in the opposite direction. One more hour later, Cindy also passes the same car. Find the speed at which Cindy drives her car.

Answer

39 mph

Solution

Alice travels 60 mph, so she passes the car after she travels

$$60 \times 6 = 360$$

miles. Similarly, Bob passes the car after traveling

$$48 \times 7 = 336$$

miles. Therefore, the car moving the opposite direction traveled

$$360 - 336 = 24$$

miles in one hour, hence is traveling 24 mph. Therefore, after another hour, the car traveling the opposite direction will be

$$336 - 24 = 312$$

miles from site A when Cindy passes it. Since Cindy has been traveling 8 hours when this happens, Cindy's speed is

$$312 \div 8 = 39$$

mph.

8   Solutions to Chapter 8 Examples

Problem 8.1 Introductory Work Questions

(a) Samantha invites Jo to help with her baking. Samantha usually prepares the pretzels for 40 minutes before putting them to the oven. Today she has Jo for help, so they worked together to prepare the pretzels. Being less experienced, Jo would have taken 60 minutes to prepare the pretzels all by herself. Working together, how long would it take for them to prepare the pretzels? In addition, it takes 15 minutes for the prepared pretzels to be baked. What is the total time needed for Samantha and Jo to prepare and bake the pretzels together?

Answer

Prep: 24 minutes, Total: 39 minutes

Solution

To answer these questions, it is not necessary to know how many pretzels they are making. We simply use the quantity 1 to represent the whole task of preparation before baking. It takes Samantha 40 minutes to complete, therefore Samantha prepares $\frac{1}{40}$ of the task in each minute. For Jo, she does $\frac{1}{60}$ of the task per minute. Working together, they can complete

$$\frac{1}{40} + \frac{1}{60} = \frac{3}{120} + \frac{2}{120} = \frac{5}{120} = \frac{1}{24}$$

of the task. So working together, they can complete the preparation in 24 minutes. Because it takes an additional 15 minutes to bake the pretzels, the total time needed is $24 + 15 = 39$ minutes.

(b) Suppose Chris can paint the entire house in fourteen hours, and Bill can do it in ten hours. How long would it take the two painters working together to paint the house?

Answer

$\frac{35}{6}$

Solution

Chris can paint $\frac{1}{14}$ of the house per hour, and Bill can paint $\frac{1}{10}$ of the house per hour.

Together they can paint

$$\frac{1}{14} + \frac{1}{10} = \frac{6}{35}$$

of the house per hour. Therefore, it takes $\frac{35}{6}$ hours for them to paint the house together.

Problem 8.2 Working Together, Working Alone

(a) Tom and Jerry can finish organizing the books at school's library together in 5 hours. If Tom do it alone, it will take him 8 hours. How long would it take Jerry to finish the same task alone?

Answer

$\frac{40}{3}$ hours

Solution

Tom can organize $\frac{1}{8}$ of the book per hour. When working together, they can organize $\frac{1}{5}$ of the book per hour. Therefore, we can find the amount of work Jerry can do alone per hour

$$\frac{1}{5} - \frac{1}{8} = \frac{3}{40}.$$

Since Jerry can organize $\frac{3}{40}$ of the book per hour, it would take him $\frac{40}{3}$ hours to finish the same task alone.

(b) Stan can load the truck in 40 minutes. If I help him, it takes us 15 minutes. How long will it take me alone?

Answer

24 minutes

Solution

Stan can load $\frac{1}{40}$ of the truck per minute. When we work together, we can load $\frac{1}{15}$ of the truck per minute. Then, we can find the amount of work that I can do per minute

$$\frac{1}{15} - \frac{1}{40} = \frac{1}{24}.$$

Since I can load $\dfrac{1}{24}$ of the truck per minute, it would take me 24 minutes to finish it alone.

Problem 8.3 Patty and Tracy can finish decorating a house for the holidays in 2.5 hours if they work together. Patty works twice as fast as Tracy. How long would it take to each of them if they work alone?

Answer

Patty: $\dfrac{15}{4}$ hours, Tracy: $\dfrac{15}{2}$ hours

Solution

Since Patty works twice as fast as Tracy (in a ratio of 2 : 1) the amount of the house they have each decorated at the end of 2.5 hours is in ratio 2 : 1 as well. Therefore, Patty has decorated $\dfrac{2}{3}$ of the house while Tracy has decorated $\dfrac{1}{3}$ in

$$2.5 = \frac{5}{2}$$

hours. Hence, Patty can decorate

$$\frac{2}{3} \div \frac{5}{2} = \frac{4}{15}$$

of the house per hour, while Tracy can decorate

$$\frac{1}{3} \div \frac{5}{2} = \frac{2}{15}$$

of the house per hour. Thus, it takes Patty $\dfrac{15}{4}$ hours to decorate the entire house and Tracy $\frac{15}{2}$ hours.

Problem 8.4 Phil can paint the garage in 12 hours and Rick can do it in 10 hours. They work together for 3 hours. How long will it take Rick to finish the job alone?

Answer

$\dfrac{9}{2}$ hours

Solution

Phil can paint $\frac{1}{12}$ of the garage in 1 hour. Rick can paint $\frac{1}{10}$ of the garage in 1 hour. Together they can thus paint

$$\frac{1}{12} + \frac{1}{10} = \frac{11}{60}$$

of the garage in an hour. After two hours, they have painted

$$3 \times \frac{11}{60} = \frac{11}{20}$$

of the garage, so $\frac{9}{20}$ of it remains. It takes Rick

$$\frac{9}{20} \div \frac{1}{10} = \frac{9}{2}$$

hours to finish painting the garage by himself.

Problem 8.5 Larger Groups

(a) James, Patty, and Joseph can organize the new products in the warehouse in 2 hours. If James does the job alone, he can finish in 5 hours. If Patty does the job alone, she can finish it in 6 hours. How long will it take for Joseph to finish the job alone?

Answer

$\frac{15}{2}$ hours

Solution

James can finish $\frac{1}{5}$ of the job himself in an hour, and Patty can finish $\frac{1}{6}$ of the job herself in an hour. Hence, working together for 2 hours they can complete

$$2 \times \left(\frac{1}{5} + \frac{1}{6} \right) = \frac{11}{15}$$

of the job. In the two hours all three worked together, Joseph must have completed the remaining $\frac{4}{15}$ of the job. Joseph can therefore complete

$$\frac{4}{15} \div 2 = \frac{2}{15}$$

of the organizing himself on one hour. Thus, it would take him $\frac{15}{2}$ hours to complete the entire job alone.

(b) Mona can complete a task alone in 150 minutes. Sarah can finish the same task in 3 hours. They work together for 30 minutes, and then a new worker, Li, joins them, and they finish the task 30 minute later. How long would it take Li to finish the task alone?

Answer

$\dfrac{225}{2}$ minutes

Solution

We know that working alone Mona can complete $\dfrac{1}{150}$ of the task in one minute. Since 3 hours is 180 minutes, Sarah can finish $\dfrac{1}{180}$ of the task in one minute. In the question above, Mona and Sarah each work a total of $30 + 30 = 60$ minutes. Therefore, Mona completes

$$60 \times \frac{1}{150} = \frac{2}{5}$$

of the task and Sarah completes

$$60 \times \frac{1}{180} = \frac{1}{3}$$

of the task. Therefore, Li, who works a total of 30 minutes, completes

$$1 - \frac{2}{5} - \frac{1}{3} = \frac{4}{15}$$

of the task, so she can complete

$$\frac{4}{15} \div 30 = \frac{2}{225}$$

of the task in one minute. Hence Li would take $\dfrac{225}{2}$ minutes to complete the task herself.

Problem 8.6 Anthony can cut a lawn in 2 hours, Mia can cut the same lawn in 3 hours, and Dandria can cut the same lawn in 2 hours. Anthony cuts the lawn for $\dfrac{1}{2}$ hour, and then Mia replaces Anthony and cuts the lawn for 1 hour herself. How many additional minutes will it take Dandria to finish cutting the lawn by herself?

Answer

50

Solution

Anthony can cut $\frac{1}{2}$ of the lawn in one hour, Mia can cut $\frac{1}{3}$ of the lawn in one hour, and Dandria can also cut $\frac{1}{2}$ of the lawn in one hour. Therefore, if Anthony cuts the lawn for $\frac{1}{2}$ hour, he cuts

$$\frac{1}{2} \times \frac{1}{2} = \frac{1}{4}$$

of the lawn. Since Mia can cut $\frac{1}{3}$ of the lawn in an hour, Anthony and Mia cut

$$\frac{1}{4} + \frac{1}{3} = \frac{7}{12}$$

of the lawn. Dandria will then need to mow $\frac{5}{12}$ of the lawn, which takes her

$$\frac{5}{12} \div \frac{1}{2} = \frac{5}{6}$$

of an hour, or 50 minutes, to finish.

Problem 8.7 Suppose if Ethan works for 5 days and Owen for 6 days, they finish a project. Alternatively, if Ethan works for 7 days and Owen for 2 days, they also finish the project. How long does it take for Ethan alone to complete the project? For Owen alone?

Answer

Ethan: 8 days, Owen: 16 days

Solution 1

Since Ethan working 7 days and Owen working 2 days is enough to complete the project, if Ethan works $7 \times 3 = 21$ days and Owen works $2 \times 3 = 6$ days, they could complete 3 projects. We also know that if Ethan works 5 days and Owen works 6 days they can complete one project. Since Owen works for 6 days in both cases, in $21 - 5 = 16$ days working alone, Ethan can finish 2 projects. Therefore, it takes Ethan 8 days to finish one project himself. Thus, in 7 days Ethan can finish

$$7 \div 8 = \frac{7}{8}$$

of the project. Hence, Owen can finish $\frac{1}{8}$ of the project in 2 days. Thus it takes Owen

$$2 \div \frac{1}{8} = 16$$

to finish a project himself.

Solution 2

Let x be the work rate of Ethan and y the work rate for Owen. If Ethan works for 5 days and Owen for 6 days, they finish the project so

$$5 \times x + 6 \times y = 1.$$

They also finish the project if Ethan works for 7 days and Owen works 2 days so

$$7 \times x + 2 \times y = 1$$

This gives us the system of equations

$$\begin{cases} 5x + 6y &= 1, \\ 7x + 2y &= 1. \end{cases}$$

Multiplying the second equation by 3, we have

$$21x + 6y = 3.$$

Subtracting the first equation from this we have

$$16x = 2,$$

so

$$x = \frac{1}{8}.$$

Substituting back into the first equation,

$$5 \times \frac{1}{8} + 6y = 1,$$

so we can solve for y, and

$$y = \frac{1}{16}.$$

Therefore, Ethan can complete the project in 8 days himself, while it takes Owen 16 days to complete the project working alone.

Problem 8.8 Alternating Work

(a) It takes Elizabeth 9 hours to complete a project. For Tiffany, it takes 12 hours. If they take turns, starting with Elizabeth, each working for one hour at a time, how much total time does it take for them to complete the project?

Answer

$\dfrac{41}{4}$ hours

Solution

Elizabeth can finish $\dfrac{1}{9}$ of the project in an hour, while Tiffany can finish $\dfrac{1}{12}$. Therefore, in a two hour time period (each working an hour) they can finish

$$\frac{1}{9}+\frac{1}{12}=\frac{7}{36}$$

of the project. Note that $36 \div 7 = 5$ remainder 1, so after 5 such periods (10 hours in total),

$$5\times\frac{7}{36}=\frac{35}{36}$$

of the project is completed, so $\dfrac{1}{36}$ remains. Since

$$\frac{1}{36}<\frac{1}{9}$$

Elizabeth can finish the rest of the project in less than an hour. Precisely, it takes her

$$\frac{1}{36}\div\frac{1}{9}=\frac{1}{4}$$

of an hour. In total, it takes Elizabeth and Tiffany

$$5\times 2+\frac{1}{4}=10\frac{1}{4}=\frac{41}{4}$$

hours.

(b) It takes Rianna 24 days to finish a job. For Helen, it takes 32 days. Rianna works on it for some days before Helen takes it over, and it takes a combined total of 26 days to get the job done. How many days does Rianna work on the job?

Answer

18

Solution 1

Rianna can finish $\frac{1}{24}$ of a job in one day, while Helen can finish $\frac{1}{32}$ of a job in one day. The target rate is 1 job in 26 days, or $\frac{1}{26}$ of a job per day. Since this is closer to how much Rianna can work in a day, we know that Rianna will work more days than Helen. We compare the differences of the rates of Rianna and Helen to the target rate. Rianna works

$$\frac{1}{24} - \frac{1}{26} = \frac{1}{312}$$

jobs per day faster than our target, while Helen works

$$\frac{1}{26} - \frac{1}{32} = \frac{3}{416}$$

jobs per day slower. The amount of days Rianna and Helen work will be in the opposite ratio

$$\frac{3}{416} : \frac{1}{312} = 9 : 4.$$

Hence, Rianna works

$$26 \times \frac{9}{13} = 18$$

days (so Helen works the remaining 8).

Solution 2

(Algebra) Rianna can finish $\frac{1}{24}$ of a job in one day, while Helen can finish $\frac{1}{32}$ of a job in one day. Let x denote the number of days Rianna works, so $26 - x$ is the number of days Helen works. Since they finish the entire job in these 26 days, we have

$$1 = x \times \frac{1}{24} + (26 - x) \times \frac{1}{32}.$$

Solving the equation,

$$x = 18$$

so Rianna works 18 days and Helen works 8 days.

Problem 8.9 Niki always leaves her cell phone on. If her cell phone is on but she is not actually using it, the battery will last for 24 hours. If she is using it constantly, the battery will last for only 3 hours. Since the last recharge, her phone has been on a total of 10 hours, and during that time she has used it constantly for 60 minutes. If she doesn't talk any more (but leaves the phone on), how many more hours will the battery last?

Answer

7 hours

Solution 1

If the phone is on but not in use, it uses $\dfrac{1}{24}$ of the battery per hour. If it is in use it uses $\dfrac{1}{3}$ of the battery per hour. Since the phone has been on but not in use for 9 hours and in use for 1 hour, Niki has used

$$9 \times \frac{1}{24} + 1 \times \frac{1}{3} = \frac{17}{24}$$

of the battery. She therefore has $\dfrac{7}{24}$ of the battery left, which will last

$$\frac{7}{24} \div \frac{1}{24} = 7$$

more hours if the phone is on but not in use.

Solution 2

Since the battery can last for 3 hours if she uses it constantly and 24 hours without using it, using the phone for 1 hour is equivalent to 8 hours without using it. Therefore, as the phone has been one for 9 hours and in use for 1 hour, we can view the phone as being on for $9 + 8 = 17$ hours. If she doesn't talk any more the battery will last

$$24 - 17 = 7$$

more hours.

Problem 8.10 Students from the Key Club at Whitman High School wash cars from the two parking lots A and B. There are four times as many cars in lot A than in lot B. First, they wash the cars in lot A for half a day. Next, half of students continue, and half of students start to wash cars in lot B. The work in lot B is done at the end of the day. Unfortunately, there are still cars unwashed in lot A. If all the students worked together to finish the cars in lot A, how long would it take?

Answer

$\dfrac{1}{4}$ of a day

Solution

(Algebra) Let x be the fraction of the total cars that the whole collection of students can wash in a full day. Since the number of cars in the two lots is in ratio $4 : 1$, $\frac{4}{5}$ of the cars are in lot A and $\frac{1}{5}$ of the cars are in lot B. Since $\frac{1}{2}$ of the students can wash the cars in lot B (which is $\frac{1}{5}$ of the total cars) in $\frac{1}{2}$ of the day, we have

$$\frac{x}{2} \times \frac{1}{2} = \frac{1}{5}.$$

Solving for x,

$$x = \frac{4}{5},$$

so all the students working together can walk $\frac{4}{5}$ of the total cards in both lots. Since all the students worked together in the morning in lot A and half of them continued in the afternoon the amount of cars washed in lot A in total was

$$x \times \frac{1}{2} + \frac{x}{2} \times \frac{1}{2} = \frac{3}{4} \times x = \frac{3}{4} \times \frac{4}{5} = \frac{3}{5}$$

of the total cars. Since lot A has $\frac{4}{5}$ of the total cars, $\frac{1}{5}$ of them remain. It will therefore take the students

$$\frac{1}{5} \div x = \frac{1}{5} \div \frac{4}{5} = \frac{1}{4}$$

of an extra day to finish the cars in lot A together.

9 Solutions to Chapter 9 Examples

Problem 9.1 The community swimming pool held 18000 gallons of water when it was full. One day, Samantha went swimming but found that the pool was being drained for cleaning.

(a) If the draining pump drained the water at the rate of 12 gallons per minute, how long would it take for the pool to be completely drained?

Answer

25 hours

Solution

The pool is drained at a rate of 12 gallons per minute, so it will take

$$18000 \div 12 = 1500 \text{ minutes, or } 1500 \div 60 = 25 \text{ hours}$$

to drain the pool.

(b) After the pool was drained, another water pump was used to fill the pool with fresh water. If the rate of filling the pool was 60 gallons per minute, how long would it take for the pool to be refilled with fresh water?

Answer

5 hours

Solution

The pool is filled at a rate of 60 gallons per minute, so it will take

$$18000 \div 60 = 300 \text{ minutes, or } 300 \div 60 = 5 \text{ hours}$$

to refill the pool.

Problem 9.2 Filling and Draining Practice

(a) There is a 10,000 liter swimming pool in Lance's community. The pool has two pipes: A and B. Pipe B delivers 1,000 liters water per hour. When pipe A and B are both on, the pool can be filled in 4 hours. How many liters per hour can pipe A deliver?

Answer

1500 liters

Solution

When pipe A and pipe B are both on, they can fill the pool in 4 hours, which means they deliver
$$10000 \div 4 = 2500$$
liters of water per hour. Then, we can find the amount of water that pipe A can deliver per hour
$$2500 - 1000 = 1500 \text{ liters.}$$

(b) A pool has two inlet pumps A and B. If pump A alone is open, it takes 12 hours to fill the pool with water. If pump B is open, it takes 18 hours to fill the pool with water. If the pool needs to be filled in 10 hours, what is the least amount of time both pumps need to be open?

Answer

3 hours

Solution 1

Pump A fills $\frac{1}{12}$ of the pool in one hour, while pump B fills $\frac{1}{18}$ of the pool in one hour. Therefore, the pumps working together fill
$$\frac{1}{12} + \frac{1}{18} = \frac{5}{36}$$
of the pool in one hour. Hence, the two pumps could fill
$$\frac{5}{36} \times 10 = \frac{50}{36}$$
of the pool in 10 hours. Since this overfills the pool by $\frac{7}{18}$, we can turn of one of the pumps for some time. If we want to have both pumps on together for the smallest possible amount of time, we turn off the slower pump (pump B). It takes pump B
$$\frac{7}{18} \div \frac{1}{18} = 7$$

hours to account for the overfilled water, so we can turn pump B off for up to 7 hours. Therefore, the two pumps must work together for at least $10 - 7 = 3$ hours.

Solution 2

(Algebra) Pump A fills $\frac{1}{12}$ of the pool in one hour, while pump B fills $\frac{1}{18}$ of the pool in one hour. Since pump A is faster, the least amount of time the two pumps work together will occur when either both pumps or just pump A are working at all times. Let x denote the amount of time both pumps work together, so $10 - x$ is the amount of time only pump A is working. Since the two pumps can fill

$$\frac{1}{12} + \frac{1}{18} = \frac{5}{36}$$

of the pool in one hour, if the pool will be filled in 10 hours we have

$$\frac{5}{36} \times x + \frac{1}{12} \times (10 - x) = 1.$$

Solving for x,

$$x = 3$$

so the pumps must work together at least 3 hours.

Problem 9.3 More Pumps and Drains

(a) There is a drain and a hose in the pool. It is known that the hose can fill the pool in 21 hours, and the drain can empty the pool in 24 hours. How long does it take to fill the pool if the drain is open at the same time?

Answer

168 hours

Solution 1

The hose fills the pool at a rate of $\frac{1}{21}$ of the pool per hour while the drain empties the pool at a rate of $\frac{1}{24}$ of the pool per hour. Since the hose is faster than the drain, if the hose is on and the drain is open, the pool is being filled at a rate of

$$\frac{1}{21} - \frac{1}{24} = \frac{1}{168}$$

of the pool per hour. It will then take 168 hours to fill the pool.

Solution 2

The hose fills the pool 3 hours faster than the drain empties it. Therefore, after filling a pool

$$21 \div 3 = 7$$

times, the hose can fill one more pool than the drain can empty in the same time. Hence, after $7 \times 24 = 168$ hours, the pool will be full if the hose and drain are open at the same time.

(b) If pump A is used alone, it takes 6 hours to fill the pool. Pump B takes 8 hours alone to fill the same pool. Uncle Sam wants to use three pumps: A, B and C to fill the pool in 2 hours. What should be the rate of pump C in order to accomplish Uncle Sam's goal?

Answer

$\dfrac{5}{24}$ of the pool per hour

Solution 1

The rate of pump A is $\dfrac{1}{6}$ of the pool per hour and the rate of pump B is $\dfrac{1}{8}$ of the pool per hour. Therefore, in two hours, pump A fills

$$2 \times \frac{1}{6} = \frac{1}{3}$$

of the pool and pump B fills

$$2 \times \frac{1}{8} = \frac{1}{4}$$

of the pool. Hence pump C must fill

$$1 - \frac{1}{3} - \frac{1}{4} = \frac{5}{12}$$

of the pool in two hours. Hence pump C must work at a rate of

$$\frac{5}{12} \div 2 = \frac{5}{24}$$

of the pool per hour.

Solution 2

(Algebra) The rate of pump A is $\frac{1}{6}$ of the pool per hour and the rate of pump B is $\frac{1}{8}$ of the pool per hour. If x is the rate of pump C, the the three pumps working together fill

$$\frac{1}{6} + \frac{1}{8} + x = \frac{7}{24} + x$$

of the pool per hour. Therefore, if the pumps fill the entire pool in two hours, we have

$$1 = 2 \times \left(\frac{7}{24} + x\right) = \frac{7}{12} + 2x.$$

Solving for x,

$$x = \frac{5}{24},$$

where x is the rate of pump C as a fraction of the pool per hour.

Problem 9.4 When two teams A and B work together, it takes 18 days to get a job completed. After team A works for 3 days, and team B works for 4 days, only $\frac{1}{5}$ of the job is done. How long does it take for team A alone to complete the job? For team B alone?

Answer

A: 45 days, B: 30 days

Solution 1

We know that if team A works for 3 days and team B works for 4 days they can finish $\frac{1}{5}$ of the job. If they work 6 times longer, team A works for 18 days and team B works for 24 days and they complete $\frac{6}{5}$ of a job. Combined with the fact that the two teams can complete the job in 18 days, we have that if team B works $24 - 18 = 6$ days it can complete $\frac{6}{5} - 1 = \frac{1}{5}$ of a job. Thus team B can complete

$$\frac{1}{5} \div 6 = \frac{1}{30}$$

of a job in one day. In 18 days team B completes

$$18 \times \frac{1}{30} = \frac{3}{5}$$

of a job, so team A must complete $\dfrac{2}{5}$ of a job in 18 days, and

$$\frac{2}{5} \div 18 = \frac{1}{45}$$

of a job in one day. We then know it takes team A 45 days and team B 30 days to complete the job if they work alone.

Solution 2

(Algebra) Let x denote the amount of work team A can complete in one day and y the amount of work team B can complete in one day. Working together they can complete the job in 18 days so

$$18 \times x + 18 \times y = 1.$$

If team A works for 3 and team B works for 4 days, $\dfrac{1}{5}$ of the job is completed, so

$$3 \times x + 4 \times y = \frac{1}{5}.$$

This gives us the system of equations

$$\begin{cases} 18x + 18y &=& 1, \\ 3x + 4y &=& \frac{1}{5}. \end{cases}$$

Multiplying the second equation by 6, we get

$$18x + 24y = \frac{6}{5}.$$

Subtracting the first equation from this equation we have

$$6y = \frac{1}{5}.$$

so

$$y = \frac{1}{30}.$$

Substituting back into the first equation,

$$18x + 18 \times \frac{1}{30} = 1,$$

so we can solve for x to get

$$x = \frac{1}{45}.$$

Therefore it takes team A 45 days working alone and team B 30 days working alone to complete the job.

Problem 9.5 Brandon, Richard, and Samuel are friends. They build a wall divider in a yard. Brandon and Richard work together, and they build $\frac{1}{3}$ of the divider in 5 days. Next, Richard and Samuel work together, and they build $\frac{1}{4}$ of the rest of the divider in 2 days. Last, Brandon and Samuel work together, they finish the rest in 5 days. How long does it take for Brandon alone to build the divider? For Richard alone? For Samuel alone?

Answer

Brandon: 24 days, Richard: 40 days, Samuel: $\frac{120}{7}$ days.

Solution 1

Brandon and Richard can build $\frac{1}{3}$ of the wall divider in 5 days, so they can build

$$\frac{1}{3} \div 5 = \frac{1}{15}$$

of the divider in one day. There is $\frac{2}{3}$ of the divider left, so in 2 days Richard and Samuel build

$$\frac{1}{4} \times \frac{2}{3} = \frac{1}{6}$$

of the divider. In one day Richard and Samuel can therefore build

$$\frac{1}{6} \div 2 = \frac{1}{12}$$

of the dividier. As

$$\frac{1}{3} + \frac{1}{6} = \frac{1}{2}$$

Brandon and Samuel build $\frac{1}{2}$ of the divider in 5 days, so they can build

$$\frac{1}{2} \div 5 = \frac{1}{10}$$

of the divider in one day. Note these 3 pieces of information tell us that if Brandon, Richard, and Samuel each work for 2 days, all together they can build

$$\frac{1}{15} + \frac{1}{12} + \frac{1}{10} = \frac{1}{4}$$

of the divider. Therefore, in one day working all together they can build

$$\frac{1}{4} \div 2 = \frac{1}{8}$$

of the divider. Recalling that Brandon and Richard can build $\frac{1}{15}$ of the divider in one day, we have the Samuel himself can build

$$\frac{1}{8} - \frac{1}{15} = \frac{7}{120}$$

of the divider in one day. Similarly, Richard can build

$$\frac{1}{8} - \frac{1}{10} = \frac{1}{40}$$

of the divider in one day. Lastly, Brandon can build

$$\frac{1}{8} - \frac{1}{12} = \frac{1}{24}$$

of the divider in one day. Therefore, it takes Brandon 24 days, Richard 40 days, and Samuel $\frac{120}{7} \approx 17.14$ days to build the divider alone.

Solution 2

Let x be how much of the divider Brandon can build in one day, y be how much Richard can build in one day, and z be how much Samuel can build in one day. Since Brandon and Richard can build $\frac{1}{3}$ of the divider in 5 days, we have

$$5x + 5y = \frac{1}{3}.$$

There is $\frac{2}{3}$ of the divider left, so in 2 days Richard and Samuel build

$$\frac{1}{4} \times \frac{2}{3} = \frac{1}{6}$$

of the divider and therefore

$$2y + 2z = \frac{1}{6}.$$

Since

$$\frac{1}{3} + \frac{1}{6} = \frac{1}{2}$$

Brandon and Samuel build the second half of the divider in 5 days, so

$$5x + 5z = \frac{1}{2}.$$

We have the system of equations

$$\begin{cases} 5x + 5y &= \dfrac{1}{3}, \\ 2y + 2z &= \dfrac{1}{6}, \\ 5x + 5z &= \dfrac{1}{2}. \end{cases}$$

Dividing the equations by 5, 2, and 5, we get a new system

$$\begin{cases} x + y &= \dfrac{1}{15}, \\ y + z &= \dfrac{1}{12}, \\ x + z &= \dfrac{1}{10}. \end{cases}$$

Adding all three up and dividing by 2, we get

$$x + y + z = \left(\frac{1}{15} + \frac{1}{12} + \frac{1}{10} \right) \div 2 = \frac{1}{8}.$$

We can subtract the 3 equations the second system of equations above from this one to solve for z, x, and y,

$$x = \frac{1}{8} - \frac{1}{12} = \frac{1}{24}, \qquad y = \frac{1}{8} - \frac{1}{10} = \frac{1}{40}, \qquad z = \frac{1}{8} - \frac{1}{15} = \frac{7}{120}.$$

Therefore it takes Brandon 24 days to build the divider himself, Richard 40 days to build the divider himself, and Samuel $\dfrac{120}{7} \approx 17.14$ days to build the divider himself.

Problem 9.6 Calvin and Tony worked on producing a set of machines. Calvin planned to complete $\frac{7}{12}$ of the task. After he finished, he helped Tony to produce 24 pieces. The ratio of the number of pieces of Calvin to Tony is 5:3. How many pieces did Tony make?

Answer

216

Solution 1

(Algebra) Suppose there are $8x$ pieces in total, so Calvin makes $5x$ pieces and Tony makes $3x$ pieces. We also know that Calvin makes $\dfrac{7}{12}$ of the total pieces plus 24 extra,

---

giving the equation so we also know that Calvin makes

$$\frac{7}{12} \times 8x + 24 = \frac{14x}{3} + 24$$

pieces in total. This gives us the equation

$$\frac{14x}{3} + 24 = 5x$$

so solving for x we get

$$x = 72.$$

Therefore, Tony makes $3x = 3 \times 72 = 216$ pieces in total.

Solution 2

Calvin originally planned to make $\dfrac{7}{12}$ of the total pieces. However, he ended up making $\frac{5}{8}$ of the total pieces. Therefore he made an extra

$$\frac{5}{8} - \frac{7}{12} = \frac{1}{24}$$

of the total pieces. Since we know this extra amounts to 24 pieces, there are

$$24 \div \frac{1}{24} = 576$$

pieces made in total. Since Tony makes $\dfrac{3}{8}$ of the total pieces, he makes

$$576 \times \frac{3}{8} = 216$$

pieces in total.

Problem 9.7 Ratios with Work

(a) It takes $\frac{1}{3}$ more time for Andy to plant one tree than for Nathan. If Andy and Nathan work together, then in the end Nathan plants 36 more trees than Andy does. How many trees are in total?

Answer

252

Solution 1

Andy takes $\frac{1}{3}$ more time to plant a tree. Since no time is mentioned at all in the problem, we may assume it takes Nathan 3 minutes to plant a tree, so it takes 4 minutes for Andy to plant a tree. Therefore, in 12 minutes, Nathan can plant 1 more tree than Andy (as Andy can plant 3 trees and Nathan can plant 4 trees). Therefore, in

$$12 \times 36 = 432$$

minutes, Nathan plants 36 more trees than Andy. In this time Nathan plants

$$4 \times 36 = 144$$

trees and Andy plants

$$3 \times 36 = 108$$

trees. A total of $144 + 108 = 252$ trees are planted.

Solution 2

(Algebra) Since it takes Andy $\frac{1}{3}$ more time to plant a tree than Nathan, the ratio of the time it takes them to plant trees is $4 : 3$. Therefore, the amount of trees they plant is in ratio $3 : 4$. Let x be such that Andy plants $3x$ trees and Nathan plants $4x$ trees. Since Nathan plants 36 more trees we have

$$3x + 36 = 4x.$$

Solving for x,

$$x = 36.$$

Therefore, there are $7x = 7 \times 36 = 252$ trees planted in total.

(b) Carolyn reads a book. Initially, the number of pages that she had read and the number of pages that she has not read are in ratio $3 : 4$. After she reads an additional 33 pages of the book, the ratio becomes $5 : 3$. How many pages does the whole book have?

Answer

168

Solution 1

At the start, Carolyn has read $\dfrac{3}{7}$ of the book. After reading 33 additional pages, she has read $\dfrac{5}{8}$ of the book. Hence, the 33 pages account for

$$\frac{5}{8} - \frac{3}{7} = \frac{11}{56}$$

of the book. Therefore, the book has

$$33 \div \frac{11}{56} = 168$$

pages in total.

Solution 2

(Algebra) At the start, the number of pages Carolyn has read and not read are in ratio $3 : 4$, so assume she has read $3x$ pages and not read $4x$ pages (so the book has $3x + 4x = 7x$ total pages). After reading 33 more pages, she has read $3x + 33$ and not read $4x - 33$. Since this is in ratio $5 : 3$ we have

$$\frac{3x + 33}{4x - 33} = \frac{5}{3}.$$

Solving for x,

$$x = 24.$$

Therefore the book has $7x = 7 \times 24 = 168$ pages.

Problem 9.8 Parts and Pieces

(a) Robot A can produce 48 pieces of machinery per hour, and robot B can produce 36 pieces per hour. After they worked together for 8 hours, there were 64 pieces marked as defective when the company tested them. How many non-defective pieces they can produce together in one hour?

Answer

76

Solution

Working together, robots A and B produce $48 + 36 = 84$ pieces per hour. However, since 64 pieces are marked defective in 8 hours,

$$64 \div 8 = 8$$

pieces are marked defective in 1 hour. Therefore, together they produce $84 - 8 = 76$ non-defective pieces per hour.

(b) A certain number of small parts need to be produced. 30 parts are scheduled to be produced after each day. After $\frac{1}{3}$ of the parts are produced, the rate of production increases by 10% thanks to improvement in efficiency. It takes 4 fewer days to produce all the parts than scheduled. How many parts are in total?

Answer

1980

Solution 1

Originally 30 parts are scheduled to be produced each day. A 10% increase in efficiency means that

$$30 \times 10\% = 30 \times 0.1 = 3$$

extra parts are produced each day. For the project to finish 4 days early, we need a total of

$$4 \times 30 = 120$$

extra parts. With the increased efficiency, this takes

$$120 \div 3 = 40$$

days. In these 40 days, a total of

$$40 \times 33 = 1320$$

parts are made. However, recall only $\frac{2}{3}$ of the parts were made with increased efficiency. Therefore,

$$1320 \div \frac{2}{3} = 1980$$

parts are produced in total.

Solution 2

(Algebra) Let x be the number of parts in total. Since originally 30 parts are scheduled to be produced each day, $\dfrac{x}{30}$ days are scheduled to produce the parts. It takes

$$\frac{x}{3} \div 30 = \frac{x}{90}$$

days to produce the first $\dfrac{1}{3}$ of the parts. With a 10% improvement in efficiency,

$$30 \times (1 + 10\%) = 30 \times 1.1 = 33$$

parts can be produced each day. It thus takes

$$\frac{2x}{3} \div 33 = \frac{2x}{99}$$

days to produce the remaining $\dfrac{2}{3}$ of the parts. Since this takes 4 days less than originally scheduled, we have the equation

$$\frac{x}{90} + \frac{2x}{99} = \frac{x}{30} - 4.$$

Solving for x,

$$x = 1980$$

so there are 1980 parts in total.

Problem 9.9 A project is planned to be completed by 45 people, and it will take some days to do it. After 6 days of work, 9 people left the team. As a result, it takes 4 more days to complete the project than originally planned. In how many days did they originally plan to finish the project?

Answer

22

Solution 1

The original team size is 45 people, after 9 leave, the smaller team to the larger team has ratio

$$36 : 45 = 4 : 5.$$

Therefore the ratio of time it takes the two team sizes to complete the same amount of work is $5:4$, so the smaller team takes 5 days to complete the work the full team can complete in 4. This means that every 5 days the smaller team works they fall 1 day behind schedule. Therefore in $5 \times 4 = 20$ days they fall 4 days behind schedule. Hence the smaller team takes 20 days to finish the project, for a total of $6 + 20 = 26$ days. Therefore the original plan had the project taking $26 - 4 = 22$ days.

Solution 2

(Algebra) Let x be the amount of work the full team of 45 can do in a single day. Therefore, the original plan is that the project takes $\dfrac{1}{x}$ days to complete. In the 6 days working at full strength, the team finishes $6x$ of the project. After the 9 people leave, it takes the remaining 36 people 4 extra days to complete the project, a total of

$$\frac{1}{x} - 6 + 4 = \frac{1}{x} - 2$$

days. The smaller team can complete

$$\frac{36}{45}x = \frac{4x}{5}$$

of the project in a day. Since the full team works for 6 days and the smaller team works for $\dfrac{1}{x} - 2$ days to complete the project, we have

$$6x + \left(\frac{1}{x} - 2\right) \times \frac{4x}{5} = 1.$$

Distributing the $\dfrac{4x}{5}$ we have

$$6x + \frac{4}{5} - \frac{8x}{5} = 1$$

and combining like terms gives us

$$\frac{22x}{5} = \frac{1}{5}$$

so

$$x = \frac{1}{22}.$$

Therefore the original plan had the project taking $\dfrac{1}{x} = 22$ days.

Problem 9.10 There are two pumps A and B. They are used to fill water in two pools that, if full, hold equal amount of water. The ratio of the water-pumping rate for pump A and pump B is $7 : 5$. After $2\frac{1}{3}$ hours, the water in the two pools, if combined, can exactly fill one pool. Next pump A increases the water-pumping rate by 25%, pump B reduces the water-pumping rate by 30%. After pump A fills the pool, how much longer does it take for pump B to fill the second pool? Express your answer as a mixed number.

Answer

$3\frac{1}{3}$ hours

Solution

Since the original ratio of the water-pumping rates is $7 : 5$, the amount of water each pump fills in a given amount of time is in ratio $7 : 5$. Therefore, pump A fills $\dfrac{7}{12}$ and pump B fills $\dfrac{5}{12}$ of a pool in the initial $2\dfrac{1}{3} = \dfrac{7}{3}$ hours. We can then calculate that initially pump A fills

$$\frac{7}{12} \div \frac{7}{3} = \frac{1}{4}$$

of one pool per hour and pump B fills

$$\frac{5}{12} \div \frac{7}{3} = \frac{5}{28}$$

of one pool per hour. Since pump A's rate then increases to

$$\frac{1}{4} \times (1 + 25\%) = \frac{1}{4} \times \frac{5}{4} = \frac{5}{16}$$

of a pool per hour, it takes pump A an additional

$$\frac{5}{12} \div \frac{5}{16} = \frac{4}{3}$$

hours to fill the first pool. Pump B's rate decreases to

$$\frac{5}{28} \times (1 - 30\%) = \frac{5}{28} \times \frac{7}{10} = \frac{1}{8}$$

of a pool per hour. Thus, it takes pump B an additional

$$\frac{7}{12} \div \frac{1}{8} = \frac{14}{3}$$

hours to fill the second pool. Hence pump B takes

$$\frac{14}{3} - \frac{4}{3} = \frac{10}{3}$$

longer to fill the second pool.

Made in the USA
Las Vegas, NV
24 April 2023

71040802R00120